THE BEAUTIFUL INVISIBLE

THE
BEAUTIFUL
INVISIBLE

creativity, imagination, and theoretical physics

GIOVANNI VIGNALE

OXFORD
UNIVERSITY PRESS

OXFORD

UNIVERSITY PRESS

Great Clarendon Street, Oxford OX2 6DP

Oxford University Press is a department of the University of Oxford.
It furthers the University's objective of excellence in research, scholarship,
and education by publishing worldwide in

Oxford New York

Auckland Cape Town Dar es Salaam Hong Kong Karachi
Kuala Lumpur Madrid Melbourne Mexico City Nairobi
New Delhi Shanghai Taipei Toronto

With offices in

Argentina Austria Brazil Chile Czech Republic France Greece
Guatemala Hungary Italy Japan Poland Portugal Singapore
South Korea Switzerland Thailand Turkey Ukraine Vietnam

Oxford is a registered trade mark of Oxford University Press
in the UK and in certain other countries

Published in the United States
by Oxford University Press Inc., New York

British Library Cataloguing in Publication Data

Data available

Library of Congress Cataloging in Publication Data

Data available

Typeset by SPI Publisher Services, Pondicherry, India
Printed in Great Britain
on acid-free paper by
Clays Ltd, St Ives plc

ISBN 978-0-19-957484-1

1 3 5 7 9 10 8 6 4 2

To my thorny flower

PREFACE

Writing a book looks like hard work, until you try writing a preface. Here you must explain yourself, reveal the motives that compelled you to write, justify your demands on the reader's valuable time. In short, you must talk rather than do, and, as Oscar Wilde once said (I paraphrase freely): 'Talking easier than doing? That's a gross popular error.'

So after much struggle and many false starts, I declare that, having spent most of my life studying and teaching physics, I felt the need to draw a circle, to recapitulate what I have learned so far about the spirit of my discipline, and to communicate it in simple language to a broad audience, to the best of my ability.

Is this true? Possibly. And partly. The idea was simple enough, but the book that emerged is, as you will see, quite unusual.

Physics, the basic science of nature, is the military academy of liberal arts. On the one hand, it is too rich in ideas, too loaded with philosophical content, too intertwined with the history of culture, to be considered merely a technical subject. A physics question rarely involves the details of a specific phenomenon—rather it concerns general patterns, regularities, laws. And the creation of new concepts in physics requires a power of imagination comparable with, if not superior to, that found in the most abstract arts, for example poetry. On the other hand, the rigour of its formulations, usually expressed in terse mathematical equations, lends this discipline a sternness that is hardly found in any of the liberal arts. Most people are frightened, frustrated, or even repulsed by this severe language.

At the very heart of physics lies a core of organized concepts and abstractions, usually, but not always, derived from experience. To this inner core of physics we give the name *theoretical physics*. Theoretical physics is the study of physical concepts and their relations to each other and to the natural world, just as physics in a broader sense is the study of the natural world. Theoretical physics is the main subject of this book.

It must be said immediately that the existence of an independent theoretical branch is a special feature of physics. Other sciences, especially the younger ones, are almost entirely empirical, i.e. based on direct observation of nature. This is not to say that these sciences do not have abstract concepts to organize what is known, for no thinking and no science would be possible without abstraction. But these concepts have not yet matured to the point of becoming themselves a primary source of scientific knowledge. In theoretical physics, however, it is possible to make discoveries about the real world using only pen and paper, without resorting to observation. So strong is our confidence (one might as well call it faith) in the validity of certain fundamental principles, that we can predict the existence of a new subatomic particle, a new planet, or a new galaxy, simply from the outcome of a mathematical calculation. Basically, we are betting that it's more likely that a hitherto unseen particle will show up after a careful and persistent hunt, than that the theory on which our calculations are based will turn out to be wrong. Obviously, it was not always like this, and it is not yet like this in any other science: such a level of confidence took a long time to build.

Does the conceptual structure of theoretical physics correspond to an objective reality in the world? Are there really electrons, photons, and quarks? Or are they just the names of useful abstractions, which happen to work very well for us, but could have been replaced by entirely different concepts if history and society had been different? About this thorny question there are two main schools of thought. Scientists are, by and large, realists. To them, a theory is just an attempt to seize something that exists, existed and will always exist, independent of our efforts to get hold of it. They would be deeply offended if they were told that the laws of nature exist only in their imagination. On the other side of the barricade, those from the humanities tend to be sceptical: to some extremists a scientific truth is nothing more than a social convention emerging from the competition between different points of view. As for myself, I think it is delirium to believe that our theories describe *literally* the world as it is. The success of a theory at explaining or predicting the facts in no way proves the objective reality of that theory. It simply demonstrates the power of our brain to successfully adjust to a reality on which we wish to prevail.

However, I totally reject the idea that scientific theories are just 'opinions', of the kind one might argue about in lunch-time conversations with friends and colleagues. On the contrary, these theories are the highest form of rational knowledge that is allowed to human beings. A good scientific theory is like a symbolic tale, an allegory of reality. Its characters are abstractions that may not exist in reality; yet they give us a way of thinking more deeply about reality. Like a fine work of art, the theory creates its own world: it transforms reality into something else— an illusion perhaps, but an illusion that has more value than the literal fact.

The world of a physical theory is not the world of platonic ideas, nor a world of plain facts. It is a tangential world, which makes contact with the world of facts in a limited region, but eventually flies off on an infinite plane, further and further from any observable reality. On this infinite plane we meet invisible actors ruled by invisible principles. Unlike lower forms of abstract knowledge, which lock the mind in a cage of unquestionable beliefs, theoretical knowledge is an open space—open to personal exploration. Precisely for this reason, no theory can be final. Subversion of the accepted wisdom is both welcome and guaranteed as long as there are thinking individuals.

The status of theoretical physics in the popular culture is mixed. On the one hand, thanks to the work of many excellent writers, there has been a growth of interest in issues such as the origin and the fate of the universe, the ultimate structure of matter, the emergence of order, the impact of new technologies. On the other hand, it remains very difficult to bridge the gap that separates these high-flown ideas from what is actually taught in introductory physics courses—the traditional curriculum that many students find more tiresome and dry than ever.

In this book I aim precisely at this gap. I do not attempt to dazzle with fashionable and modern ideas, nor to teach physics in the traditional way. Physics cannot be taught without mathematics and there is hardly any mathematics in this book. My objective is to engage and entertain with the elegant beauty of the ideas and methods of theoretical physics, which I believe can stand on their own, even without the allure of grand-unified theories or unlimited technological power. In particular, I wish to show how abstract theoretical ideas work their way into the mess of the material world, filling it with a meaning that does not intrinsically belong to it. These ideas are the soul of theoretical physics, and the journey I propose to undertake will be like a tour through a gallery of invisible paintings, of which we can nevertheless perceive the colours.

This book was conceived and partly written during many flights between America and Europe. So if you sense something odd about it—something like being suspended 30,000 feet above the ground in a strange place that is neither here nor there—you are not mistaken: that's precisely how it is.

Giovanni Vignale

ACKNOWLEDGEMENTS

I am very grateful to my editor, Latha Menon, for giving me the chance to publish this unusual book. Her criticism was frank, constructive, and always tactful. Besides alerting me to weaknesses and inconsistencies, she made a valiant effort to expunge political incorrectness, misplaced sentimentalism, cheap mysticism, and overly sensuous imagery. All in all, I believe she has gone a long way towards saving me from myself, or so I hope. I am deeply indebted to Phil Henderson for coming up with an elegant and catchy title, which supersedes my rather stiffer 'Praise of the Abstract'. I thank Emma Marchant and Debbie Protheroe of the OUP production team for help with the graphics, Claire Thompson for expertly supervising the production process, and Paul Beverley for his careful and critical copy-editing. Finally, it is a special pleasure to thank my daughters, Sonia and Veena, for reading parts of the manuscript at an early stage, and not mincing words in their reviews. I hope they will keep this book as a photograph of my mind as it was when I was young.

CONTENTS

Isn't that like a bridge consisting only of the first and the last pillar,
and yet you walk over it securely as though it was all there?...
But the really uncanny thing about it
is the strength that exists in such a calculation,
holding you so firmly that you land safely in the end.

Robert Musil, *The Confusions of Young Törless*

Reality is invisible

Antoine de Saint-Exupéry was a French nobleman by birth and an aeroplane pilot by choice: but by destiny he was always a writer. Plane crashes were common and quite survivable in the early days of aviation. Saint-Exupéry survived many, but none as memorable as the one that, on the night of 30 December 1935, shattered his dreams of establishing a new record of speed from Paris to Saigon. Without the global positioning system... actually without even a radio to call for help... he and his mechanic wandered in the Sahara desert for five days and nights, desperate, parched, on the verge of delirium, before being rescued by a caravan of merciful Bedouins. It was during this ordeal that the two men resolved never to turn down a drink for the rest of their lives.

Not surprisingly then, Saint-Exupéry's best-known work—a short tale entitled *The Little Prince*—begins with a pilot crash-landing in the desert. This pilot, at the age of six, had abandoned a promising career as an artist, discouraged by the grown-ups' lack of understanding. His very first drawing was seen, by the adults, as a hat—even though it clearly represented a boa digesting an elephant. This is just the first hint of what quickly emerges as the main theme of the tale—the idea that *anything essential is invisible to the eye* and that *one sees clearly only with the heart*. In the big and lonely desert the pilot meets his own self, thinly disguised as a little boy from another planet. The boy asks many questions and answers none. Following him, we get to know a moon-coloured snake who solves all riddles, a wise fox whose only wish is to be tamed, a thorny flower who coughs and shivers in the wind, a tiny planet on which the sun sets every few minutes, and, most importantly, a hidden well, the well that every desert hides in its centre, the well that makes every desert beautiful. And thus we learn that 'whether it's a house, or the stars, or the desert, what makes them beautiful is invisible'.[1]

To a religious person this message will hardly come as a surprise. The mystical authorities of all times and cultures insist on the absence of sensible images as the necessary condition for the contemplation of the higher truths. Likewise there is a consensus that true wisdom, in spiritual and in artistic matters, comes with the ability to see things that are normally hidden from sight—to see them with the heart. This is not surprising. Poetry, philosophy, religion are among the oldest expressions of human thought, and 'oldest' in this case means 'youngest', for the three of them sprang out vigorously and simultaneously in the hearts of young people who knew almost nothing about the world, but were naturally predisposed to believe in the reality of the invisible. On the other hand, few people would spontaneously associate the idea of an invisible reality to *science*. With science, it is said, human thought comes of age. We abandon the sentimental/emotional approach of childhood and enter a realm of rationality and objectivity. This tough goddess probes Nature with clocks and scales and absolutely straight

---

[1] Quite fittingly Saint-Exupéry achieved a measure of invisibility which is rarely granted to celebrities: his body disappeared in the waves of the Mediterranean sea on 31 July 1944.

rulers, not to mention the scourge of mathematical equations and formulas. In its presence, only what can be clearly defined and measured has a right to exist. She looks at us with cold, self-assured eyes. No wonder most people are unable to love her.

And yet, even cold-hearted science is not just what meets the eye. The study of nature may begin with direct observation and measurement, but in order to understand what we see (and also, somehow, to 'tame' it) we find ourselves compelled to go beyond the visible, to reach for hidden forms, for fundamental principles working behind the scenes. Francis Bacon, one of the founders of modern empirical science, wrote that 'Nature cannot be commanded except by being obeyed'. I would add that nature cannot be understood except by being transcended. Nowhere is this more evident than in that special kind of science which is called *theoretical physics*. I emphasize the word *theoretical* because, as we will soon see, this is a kind of science distinct and somewhat different from *physics*. When physicists work on a theory, they are not dealing directly with nature, but with an abstract model in which they have already decided which aspects of reality must be absolutely retained, and which ones can be dismissed. Often, in creating this model, they make bold and quite implausible assumptions, which can only be validated by the consistency of the results. But, to take such bold steps one cannot rely on calculation alone: it takes passion, imagination, a sense of beauty—all things that we grasp with our whole personality, and definitely with our heart.

This book is an invitation to approach theoretical physics from this particular angle: as the science of the invisible, as a modern form of theology. No particularly new or revolutionary ideas will be presented; nevertheless, it will be quite a journey. We will proceed through a land populated by strange characters—point particles, light rays, minimum principles, conservation laws, invisible fields that reach out to the remotest regions of the universe—all of which strongly remind us of something real, yet are nowhere to be seen; they emerge from a process of abstraction which stretches to the limit something we have long been familiar with. We will acquaint ourselves with universal laws that rule the behaviour of these characters—only to realize that even the universal laws have limits and can be transcended. We will see how every single element of this surreal landscape can be one and multiple—diverse and fixed. And, finally, at the centre of it all, we'll find the enthusiasm and the sadness of the Little Prince, smiling and waiting for us with a list of questions that can never be fully answered.

2

The way of the abstract

A science within the science

Physics, most of us would agree, is the basic science of nature. Its purpose is to discover the laws of the natural world. Do such laws exist? Well, the success of physics at identifying some of them proves, in retrospect, that they do exist. Or, at least, it proves that there are Laws of Physics, which we can safely assume to be Laws of Nature.

Granted, it may be difficult to discern this lofty purpose when all one hears in an introductory course is about flying projectiles and swinging pendulums, strings under tension and beams in equilibrium. But at the beginning of the enterprise there were some truly fundamental questions such as: the nature of matter, the character of the forces that bind it together, the origin of order, the fate of the universe. For centuries humankind had been puzzling over these questions, coming up with metaphysical and fantastic answers. And it stumbled, and it stumbled, until one day—and here I quote the great Austrian writer and ironist, Robert Musil:

> ...it did what every sensible child does after trying to walk too soon; it sat down on the ground, contacting the earth with a most dependable if not very noble part of its anatomy, in short, that part on which one sits. The amazing thing is that the earth showed itself uncommonly receptive, and ever since that moment of contact has allowed men to entice inventions, conveniences, and discoveries out of it in quantities bordering on the miraculous.[1]

This was the beginning of physics and, actually, of all science: an orgy of matter-of-factness after centuries of theology. Careful and systematic observation of reality, coupled with quantitative analysis of data and an egregious indifference to theories that could not be tested by experiment became the hallmark of every serious investigation into the nature of things.

But even as they were busy observing and experimenting, the pioneers of physics did not fail to notice a peculiar feature of their discipline. Namely, they realized that the laws of nature were best expressed in an abstract mathematical language—a language of triangles and circles and limits—which, at first sight, stood almost at odds with the touted matter-of-factness of experimental science. As time went by, it became clear that mathematics was much more than a computational tool: it had a life of its own. Things could be *discovered* by mathematics. John Adams and, independently, Urbain Le Verrier, using Newton's theory of gravity, computed the orbit of Uranus and found that it deviated from the observed one. Rather than giving up, they did another calculation showing that the orbit of Uranus could be explained if there were another planet pulling on Uranus according to Newton's law of gravity. Such a planet had never been seen, but Adams and Le Verrier told the astronomers where to look for it. And, lo and behold, the planet—Neptune—was there, waiting to be discovered. That was in 1846.

[1] Robert Musil, *The Man Without Qualities*, Chapter 72.

Even this great achievement pales in comparison with things that happened later. In the 1860s, James Clerk Maxwell trusted mathematics—and not just the results of a calculation, but the abstract structure of a set of equations—to predict the existence of electromagnetic waves. And electromagnetic waves (of which visible light is an example) were controllably produced in the lab shortly afterwards.

In the 1870s Ludwig Boltzmann undertook the task of finding out, by mathematical analysis, how a hypothetical world made of *atoms* would behave. Nobody had seen an atom, and very few believed seriously in what, at the time, must have looked like a very artificial concept. With the help of a revolutionary mathematical approach in which probability was the main actor, Boltzmann was able to show that his artificial world behaved pretty much like the real world. At least, the behaviour of gases was the same!

These three examples illustrate three different ways of practising the strange kind of science known today as *theoretical physics*. They are like three different literary genres, such as essay, poem, and novel. In the first, one applies a general theory, summarized in a set of mathematical equations, to the solution of a concrete problem. In the second, one plays with the mathematics to find new equations that are more satisfactory from an intellectual, aesthetical, or practical point of view. Finally in the third way—the Boltzmann way—one constructs an artificial world with building blocks that obey the laws of a previously established theory. Then one tries to find out whether the behaviour of this artificial world matches the behaviour of the real one.

By the early twentieth century theoretical physics had become a well—established science within the science. The two great triumphs of that period—relativity and quantum mechanics—spawned a host of revolutionary concepts such as 'antimatter' and 'black holes', which were discovered in later experiments and have since become staples of popular scientific literature. No other science, as far as I know, can boast a comparable record of successful predictions. And yet, at the beginning of the twenty-first century theoretical physics stands aloof in the middle

FIG. 1 René Magritte, *The Castle in the Pyrenees*. (Reproduced with permission)

of a culture that is deeply suspicious of abstract thought. Indeed, to many people a theory is the exact opposite of a science. The very word 'theory' suggests a loss of contact with reality, which in turn evokes a lack of vital strength, of feeling, of generosity—in short of all the virtues that are most prized in a human being.

I must concede that the prejudice against the abstract is not entirely unjustified. All around us we see abstract concepts, laws, and classifications prevail upon basic considerations of humanity. Governments and individuals have been known to commit the worst crimes in the name of abstract ideals. Yet I believe that the main problem in those cases is not the faith in the abstract, but a specific degeneration of the abstract, which I will describe later, and for which I suggest we use the term *formality*.

In this book I'll try to show that, contrary to prejudice, theory is one of the highest form of knowledge to which we can aspire. Building a theory is basically the same as recreating the world in a way that makes it meaningful to us on all levels (not only the rational one). It is an attempt to create a connection with a universe that would otherwise remain mysterious and indifferent. When I think of theoretical physics I do not see an edifice resting on square foundations on the ground, but a structure closed on itself like the castle of Magritte's painting (Fig. 1). At the bottom I see the heavy, rough mass of the real facts in need of explanation. At the top I see a graceful composition of roofs and turrets—the theory. Facts and theory need each other like two parts of the same body. The rock supports the castle, but the castle holds the rock and lifts it to a higher level. It looks as if the whole thing should come down in a big crash, but it doesn't. A mysterious power keeps it suspended above the waves of the ocean: it is the power of internal *consistency*.

Great art arises from a fortunate encounter of observation, imagination, and technical skill. In Musil's words 'a poem, with its mystery, cuts through the point where the meaning of the world is tied to thousands of words of constant use, severs all those strings, and turns it into a balloon floating off into space'.[2] Theoretical physics has all in common with poetry, except the lack of restraints. It is fiction constrained by fact. Unlike the balloon of Musil's metaphor, it must lift the heavy load of its own foundation—the empirical facts. Thus, the theory lives at the interface between the fictional and the real world, all along the fuzzy frontier between what can be seen and what can only be imagined.

The tapestry

An engineer turned writer, Robert Musil dreamed of applying rational methods to what he dubbed 'the non-ratioid sphere'—the realm of unmeasurable, irreproducible phenomena of the spirit. Once, in an interview, he said: 'I am not interested in the real explanation of real facts. My memory is bad. Furthermore, facts are always interchangeable. I am interested in the spiritually typical, I could even say: in the spectral dimension of what happens'.[3]

This statement may well serve as an introduction to the theorist's view of the world. I don't think that my colleagues will be offended if I say that, like Musil, we have bad memory, and intensely dislike carrying around large amounts of information. We'd rather keep in mind a few key ideas from which everything else can be worked out. The power of these ideas lies in their universality, i.e. in the fact that they unify large classes of apparently different phenomena. A classic metaphor describes these ideas as the long threads of a tapestry. 'Nature uses only the longest threads to weave her patterns, so each small piece of her fabric reveals the organization of the entire tapestry' writes Richard Feynman, one of the legendary theoretical physicists of the twentieth century.

[2] Robert Musil, *The Man Without Qualities*, Chapter 84.
[3] Robert Musil, Interview with Oskar Fontana, 30 April 1926.

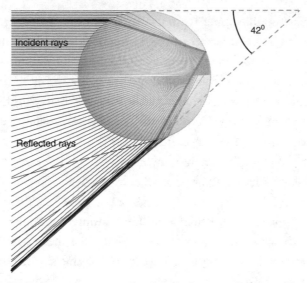

FIG. 2 Path of light rays in a water droplet. The largest reflection angle—shown by the thick black line—is about 42°. In keeping with the principle of the stationary maximum, most rays exit the droplet with an angle close to this value.

Undoubtedly, this view of the world runs counter to some of our most powerful instincts, which urge us to concentrate on what is near and present, to prefer details to generalities, to react promptly and vigorously to small changes, and to resort to abstract thinking only when everything else fails. Facts are definitely not interchangeable when your survival depends on them. On the other hand, the theoretical mind is like a glass, which splits the incoming light into its component colours. These component colours constitute the 'spectral dimension of what happens'.

Let me give you an example to which I am sure everyone can relate: the rainbow, a universal symbol of peace between Nature and human-kind. How do we understand it? At first, one might be tempted to think of it as an optical illusion. But this is not right, for this 'illusion' coincides with the observation of something quite tangible—a set of genuine water droplets which reflect the light of the sun—a fact first realized by the Persian philosopher Qutb al-Din al-Shirazi. Figure 2 shows the path of light rays coming from the sun and reflected by a droplet of water. A typical ray changes direction three times: the first time, when it enters the droplet (this change of direction is known as *refraction*); the second time, when it is reflected from the inner surface of the droplet, as if from a mirror; the third and last time, when it re-emerges in the air. Looking carefully at Fig. 2, you notice that the emerging light concentrates in a direction forming an angle of 42° with the direction of the incoming light (the exact value of this angle depends slightly on the colour of the light). There are, to be fair, a few stray rays that emerge at an angle distinctly smaller than 42°—for example, the one that hits the droplet in the centre and comes back retracing its path. But these are quite exceptional. Most rays are reflected at nearly 42°, and no ray is reflected at more than 42°.

The concentration of reflected light about the maximum angle is the consequence of an abstract mathematical principle, which plays an incredibly important role in our understanding of nature. I'll call it 'the principle of the stationary maximum', even though the same is true for the minimum, so I could as well call it 'the principle of the stationary minimum'. The principle says that the rate of variation of any variable quantity vanishes when that quantity approaches its maximum or minimum value. Think about it. When you throw a ball vertically upwards, it stops for an instant at the point of maximum height, before beginning its descent. For just a fleeting moment, it is stationary: the rate of variation of its height vanishes. When the sun in its yearly revolution reaches the northernmost limit of its path in the sky (on 21 June) it lingers there, nearly stationary, for a few days: this is the origin of the word *solstice*, which means *solar arrest*. In the present case, the angle of reflection of a light ray varies gradually as the point of entry of the ray shifts from the periphery to the edge of the droplet, but ever more slowly as it approaches the critical point for which the deflection angle has the maximum value of 42°. And this is why we observe such a large concentration of rays reflected at precisely 42°: basically, it

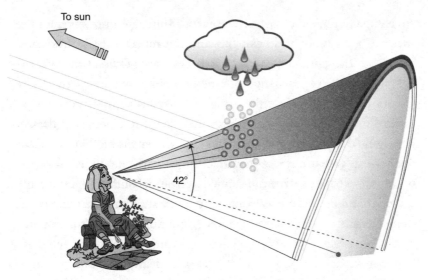

FIG. 3 The geometry of the rainbow. The three colour bands, starting from the top, are red, green, and blue.

does not matter whether the rays enter the droplet a little above or a little below this critical point: the deflection angle remains essentially unchanged.

Imagine now looking at a large collection of water droplets from far away, with the sun at your back. From your position you can only see those droplets which happen to reflect light into your eye. These droplets are easily seen (in your mind, I mean!) to lie on the surface of a cone emanating from your eye in the direction opposite to the sun. This cone is depicted in Fig. 3. Actually, because the sunlight has components of different colours, and because different colours are reflected at slightly different angles (the angle gradually decreasing as the colour changes from red to green to blue), the surface of the cone broadens into bands of different colours, with the droplets near the top appearing red, those in the middle green, and those at the bottom blue. This is also shown in Fig. 3.

I have gone into some details of the 'theory of the rainbow' just to bring out the 'spectral components' of the phenomenon. Some of these are completely abstract, like the concept of the 'light ray', which I have sneaked under your nose as if it were the most natural thing in the world. But there are no 'light rays' in nature, as every serious painter of light knows. There is light, of course, which we understand as the vibration of an abstract entity called the electromagnetic field, in itself invisible. And there is colour—a curious side—effect of the frequency of vibration of the electromagnetic field acting on the cells of the retina. But light rays are 100 per cent abstraction, arising—as we will see in one of the later chapters—from a subtle application of the principle of the stationary minimum to electromagnetic waves. Other components, like the laws of reflection and refraction of light rays, are also consequences of that principle.

According to Plato's myth of the cave, everything we perceive in this world is the imperfect shadow of a perfectly formed idea, which gloriously shines in an ideal world beyond the stars. It is called 'the myth of the cave' because the shadows appear on the end wall of a cave wherein we humans are imprisoned. The rainbow suggests a different metaphor. Our perceptions are not like the shadows of a higher reality, but like the patterns that we are able to make out on a sea of microscopic droplets—the latter being actually a lower, not a higher reality. The microscopic droplets correspond to the 'elementary constituents of matter'. On that background shine the rainbows of our perceptions, and on the background of our perceptions shine the rainbows of our theories, which, like all rainbows, are visible only from the right distance and at the proper angle.

Fantastic precision

Building theories is not as easy as it might sound. One must walk a fine line between dismissing a huge amount of detailed information (the 'real facts') and intensifying the 'spectral dimension of what happens' (Musil). Discarding information is as important as gathering it. Discard too much, and you are left with a soulless abstraction. Discard too little and you are trapped in a labyrinth of complexity. By discarding the right amount, one creates something that is somehow more real than reality—a metaphorically exact world. 'Fantastic precision' is the name of the game. Dream up possibilities, visualize them, analyse them with honesty and courage, with all the powers of your being, rational and not. Theory grows at the confluence of fantasy and truth.

The great inventor and electrical engineer Nikola Tesla was a visionary of the highest power, who saw engines in his mind before seeing them in the lab. This is how he describes his method:

> ...When I get an idea I start at once building it up in my imagination. I change the construction, make improvements and operate the device in my mind. It is absolutely immaterial to me whether I run my turbine in thought or test it in my shop. I even notice if it is out of balance... Invariably my device works as I conceived that it should...Why should it be otherwise?[4]

The deadpan humour of this quote should not obscure the main point: visions can be wrong or out of balance, like turbines; there is precision in fantasy.

At around the same time that Tesla was dreaming about turbines, the American artist Charles R. Knight (who, incidentally, was almost completely blind) was creating the *dinosaur* out of his own imagination at least as much as from scientific studies of fossils. Figure 4 shows one of his many visions: a scene from a prehistoric world of the sea. This scene cannot be called 'real' because it includes elements such as the colour of the creatures and other details that nobody has ever observed; yet it cannot be called 'fictional' because the shape of the creatures has been inferred from painstaking studies of anatomical clues. It is both real and fictional.

The relation between fact and fiction, fantasy and truth, lies at the heart of Mikhail Bulgakov's great novel *The Master and Margarita*. The novel unfolds against the background of the atheistic, matter-of-fact

---

[4] Nikola Tesla, *My Inventions*.

FIG. 4 A real and fictional world of prehistoric animals created by Charles R. Knight. (Reproduced with permission)

culture of the Soviet Union in the 1920s. The Master is a great writer, author of a fantastic novel about Jesus and his prosecutor Pontius Pilate. There is a deliberate confusion between the two novels—the one we are actually reading and the one we are reading about. Like Bulgakov's novel, the Master's novel has been soundly thrashed by the critics. Like Bulgakov, the Master has burned the manuscript and suffered a nervous breakdown. Now languishing in a mental asylum, he learns from a fellow inmate that a certain Professor Woland—a foreign expert of black magic—has come to Moscow and is lecturing about the life of Jesus. Professor Woland claims to have been present at the interrogation of Jesus by Pilate, and his account of the facts confirms word by word the Master's manuscript. What is the Master's reaction? Anger? Surprise? Disappointment? Does he accuse Professor Woland of plagiarism? Not at all. He folds his hands as if in prayer and whispers 'Oh, I guessed right! I guessed everything right!'

One must appreciate the peculiar point of view from which the story of Jesus and Pilate is told within *The Master and Margarita*. It is not the point of view of a reporter and it is not that of an ordinary eyewitness: it is 100 per cent the point of view of the fantastic writer—a point of view that happens to coincide with that of the supernatural eyewitness, Professor Woland, alias the Devil. And it is precisely this fantastic point of view that makes the story 100 per cent true and exact, as opposed to just another newspaper report.

How does the Master manage to grasp the truth that has been murdered by politicians, muddled up by unreliable witnesses? Apparently, by dreaming it up, or, in his own words, by *guessing* it. But the dream is one of excruciating precision, an artist's via dolorosa, where every detail captures the essence of reality so accurately and truthfully that questions such as 'Did it really happen?', 'Did it really exist?' become a concern for those who have already missed the point. The Master is a great artist, but also a model for the theoretical physicist.

Fantastic precision is a difficult art. Coming up with a good theory in science is not just difficult; it is almost impossible. Yes, we are steadily bombarded by fancy ideas about angels and demons and invisible forces and fields, but most of these fantasies lack the essential feature of precision. They are cheap speculations, or remakes of ancient theories that were once alive. In fact, almost every product of our imagination is bound to be ultimately crushed by reality. One can say of theory what the anonymous author of the *Carmina Burana* said of fortune: 'she has a fine head of hair, but when it comes to seizing an opportunity, she is bald'.

This is why, whenever a real insight occurs, which seems free of contradiction as far as one can see, and pulls together many facts, and is therefore beautiful; we salute it as a momentous event. If, furthermore, the theory has predictive power—as should always be the case in science—then it is hard to avoid the conclusion that there must exist a real feature of the world—an objective reality—which is correctly described by that theory. Just as to many people the origin of life would be inexplicable without a Creator, so to most scientists the success of a theory would be inexplicable without an objective reality behind it. The theory is canonized in textbooks; it spawns churches and authorities; it dictates the correct interpretation of the facts. Time goes by, and we see the daughter of a daring imagination lose its inner fire, forget that it could have been different, wither up to dogma. But no theory is so good that it is not permanently at risk of being superseded by deeper insights and generalizations.

The formal and the abstract

Before ending this chapter and beginning our journey into the abstract, I would like to clear a confusion that often arises between two attitudes that look superficially similar, but are really as different as day and night. I am talking of the confusion between formality and abstraction. The latter is human—it has emotional content. Formality is what is left of abstraction after the emotional content has boiled away. It is the abstract reduced to convention.

At its best, a formalism is a way to organize our thinking so that we don't waste time rediscovering certain basic steps of reasoning, but focus our intellectual energy on what is truly new and challenging. It is like the automatism of the fingers allowing the pianist to concentrate on the spirit of the music. A good example in point is the use of algebra to deal with unknown quantities as if they were already known. Precise rules of manipulation allow us to do this very efficiently, without much intellectual effort. Actual numerical values are 'plugged in' only at the very last stage in the solution of a problem. This is how, for example, your banker can tell you at a moment's notice the amount of the monthly payment that will repay a debt of D pounds in M months at an interest rate of I per cent, leaving you plenty of time to wonder whether you can afford it.

Another benefit of formality is that it protects us from errors, sophisms, or outright abuses that would inevitably pollute a system in which matters of truth and justice were approached informally. Think only how easy it is, even with the best of intentions, to mix the truth and what we wish to be the truth, when we are burning with enthusiasm for a new idea. And now think what can happen when there is a deliberate intent to deceive. These are dangers against which formality offers good protection, providing an impersonal and impartial method for checking the correctness of arguments and procedures.

Early in the twentieth century mathematicians were fascinated by the idea that every true statement in mathematics could be derived logically, inescapably, from a very small set of assumed truths, known as *axioms*, through the application of strict rules of deduction. The assumed truths were to be expressed as strings of symbols, just as every sentence of this book is a string of characters of the English alphabet. The rules of deduction would tell us how to construct grammatically correct sentences from sentences that were already known to be correct. Truth would be no more and no less than grammatical correctness, and could be established without reference to meaning. Then one could go on to discover systematically, one after the other, all the truths contained in the axioms. Such an exploration could be carried out quickly and flawlessly by a machine far more efficient than any human mathematician. It would be the ultimate antidote to the fly of Blaise Pascal[5] or to the bee of Emily Dickinson[6]—those fatal distractions that interpose themselves between the mind and the truth. Such were the suicidal thoughts of early twentieth—century mathematicians.

Thank God these hopes turned out to be short-lived. In 1931 the logician Kurt Gödel announced the discovery of a dramatic theorem, according to which every formal system throws in the towel when confronted with a particular truth that cannot be proved within the system itself. The theorem is dramatic because it turns formality against herself: it uses formal analysis to expose the limitations of formal systems. In short, the theorem shows that there are things which we see to be true when we think 'out of the box' (the box being the formal system), but not when we think 'within the box', i.e. sticking to the rules of the formal system. In other words, *more things are true than are*

---

[5] 'The mind of this supreme judge of the world (man) is not so independent as to be impervious to whatever din may be going on nearby...Do not be surprised if his reasoning is not too sound at the moment, there is a fly buzzing round his ears; that is enough to render him incapable of giving good advice.' (Pascal, *Pensées*, 48 (366)).

[6] 'I heard a Fly buzz–when I died –/...I willed my Keepsakes–signed away/What portion of me be/assignable–and then it was/There interposed a Fly/With Blue–uncertain stumbling Buzz –/ Between the Light–and Me –/ And then the Windows failed–and then/ I could not see to see –' (Emily Dickinson).

formally provable.[7] I emphasize this conclusion because truth in general (regardless of our limited ability to prove it by accepted means) is the proper object of theory. Formality, on the other hand, recognizes only *provable truth*, e.g. Gödel's theorem itself. This difference implies that not all parts of a theory need to be provable in a formal sense: in fact, quite often they support each other and hold together by virtue of a global internal consistency that is somehow stronger than a proof.

Until now I have been talking about the bright side of formality, but now I must say something about its dark side. At its worst, formality is like a screen interposed between us and reality; it is a willingly embraced form of blindness.

I must immediately say that nothing could be further from my intention than to be disparaging blindness in general terms. In fact, there is a profound connection between blindness, abstract thought, poetry, and spiritual wisdom. Homer, the father of all poets, discovered his literary vocation only when he was no longer encumbered with a sense of vision. Democritus, the inventor of the atom, plucked out his eyes to think more clearly. And Chiu-Fang Kao, the most experienced of the imperial horse-trainers, could not tell whether a certain horse was a dun-coloured mare or a coal-black stallion: '...intent on the inward qualities, he loses sight of the external... He looks at the things he ought to look at, and neglects those that need not be looked at.'[8]

---

[7] I admit that this point must sound unconvincing at first. How can we confidently say that something is true if we cannot prove it? To get a taste of the argument we must imagine a proposition **G** which becomes false as soon as it is proved to be true. For example, consider the proposition **G** which says: 'It is impossible to prove that **G** is true.' This proposition **G** cannot be *proved* to be true, for, if we could do it, we would have actually proved that 'It is impossible to prove that **G** is true', contrary to what we have just done! But precisely because we can't prove it to be true, we see that proposition **G** is, in fact, true (read it again!)–but not in such a way that can be proved.

[8] From *The Book of Chuang Tzu*, as quoted by J. D. Salinger in *Raise High the Roof Beam, Carpenters*.

But, what these great seers have in common is that their blindness is blindness to the trivial and the non-essential—a status one attains (if ever) only after long study and careful observation, never before (remember: discarding information is as important as gathering it). Opposite to this, the blindness I am describing here comes *before* observation and before even knowing if there is anything worth observing. A classic example of blindness by choice is that of Queen Gandhari, wife of the blind-born King Dhritarashtra, whose clouded judgement and partial rule ushered in *Kali Yuga*—the age of downfall of the human race.[9] The Queen was not blind, and could perhaps have saved her husband and the kingdom from many disastrous blunders, had she not chosen to blindfold herself, through a misguided notion of virtue in sharing her husband's darkness.

The abuse of formality typically begins with the forcing of reality into the mould of preconceived schemes and fantastic ideas that have little in common with reality. Bureaucrats are the world champions of this attitude. Think of the difficulty of filling in a form when none of the proposed options applies. The difficulty arises because the questionnaire was not written for a real person but for an abstract average person, who exists only in the minds of the bureaucrats.

The administration of justice is another area in which one often confronts the disastrous consequences of extreme formality. One of the memorable characters of Musil's *Man Without Qualities* is Moosbrugger—the dim-witted man who lives a peaceful life until the day on which, in a bout of insanity, he murders a woman. Is he guilty or innocent? At the trial, the psychiatrists are absolutely clear: Moosbrugger's syndrome does not exactly correspond to any hitherto observed syndrome: any further conclusion is entirely left to the jurors. About this the narrator comments:

> 'The precision with which Moosbrugger's peculiar mentality was fitted into a two-thousand-year-old system of legal concepts resembled a

[9] The story is told in the classic Indian epic *Mahabharata*.

madman's pedantic insistence on trying to spear a free-flying bird with a pin...The courtroom...offered an image of life itself, in that all those energetic up-to-the-minute characters...abandon questions of beauty, justice, love, and faith...to a subspecies of men given to intoning thousand-year-old phrases about the chalice and the sword of life.[10]

The formal person, also known as a *philistine*, is he or she who chooses to honour the letter of the law rather than its spirit. The essence of the philistine's attitude is a little like the attitude that considers true only what is provable from a given set of assumptions. Genuine theory has little use for such attitude. Jesus fought against it passionately. In the Gospels his strongest words are against the Pharisees, the high priests of the law, to whom religion is nothing but form. He condemns them as 'blind guides...who have neglected the most important part of the law—justice' (Matthew 23:23). He accuses them of 'shutting the kingdom of heaven in men's faces' (Matthew 23:13), and 'taking away the key to knowledge' (Luke 11:52). With immortal metaphors (*whitewashed tombs, unmarked graves, snakes, brood of vipers*) he delivers them to infamy. But even he could not defeat them. And today, bolder than ever, they sit by the thousands in the courthouses, in the churches, in government offices—wherever power is wielded upon man.

Praise of the abstract

Like rain and sun in a day of March, the formal and the abstract alternate in the story of Eklavya and Drona—the best student and the worst teacher of all times. The boy Eklavya wishes to learn archery, but Drona, Guru to the Pandava Princes, refuses to teach him because Eklavya was born into a caste whose members are not allowed to be warriors (the formal). Undaunted, Eklavya makes a clay idol of Drona (the abstract) and, training assiduously under its spell, manages to become the greatest archer in the world. When Drona discovers this,

---

[10] Robert Musil, *The Man Without Qualities*, Chapter 62.

he becomes terribly angry, and, with a callousness that makes our private universities look like charitable institutions, demands *gurudak-shina*—yes, payment for tuition! Eklavya, who has always considered himself Drona's pupil, is overjoyed and grateful to be so recognized as his student (the formal again). Then, at Drona's request, he readily agrees to cut off the thumb of his right hand—thus putting an end to his dreams of archery.

The story has more than one lesson to teach. Eklavya rises as long as he believes in the abstract Drona of his own imagination, and falls when he becomes a student of the real Drona. In the abstract he finds inspiration to break the constraints of formality and to pursue the highest ideals. Likewise, in the abstract we learn to recognize ourselves in other human beings—not only our immediate neighbours (which is relatively easy), but everybody regardless of social status, role, race, gender, nationality, citizenship... And not because we are blind to these attributes, but because we see them very clearly as aspects of their individual being. So we resent the injustice done to others as if it were done to us personally.

To the traditional image of justice as a blindfolded goddess, armed with sword and balance and ostentatiously unaware of her subjects, I should like to oppose the exquisite *Woman Holding a Balance* by the Dutch painter Vermeer. The painting is reproduced in Fig. 5. The woman holds the balance with just three fingers. The gentle grip suits the delicate balance, which is an assay balance—a fine scientific instrument for weighing very small quantities of precious metals. It is the 'exquisite and just' balance that gives the title to one of the most famous scientific essays of all times: *The Assayer* by Galileo Galilei. The thick wooden table is cluttered with valuable merchandise: pearls, necklaces, velvets—but none of this is on the balance. The plates of the balance are empty and the woman's serene gaze rests on that emptiness. On the back wall a painting within the painting depicts a terrifying scene of the Final Judgement. The woman's head hides the place where Saint Michael would be weighing souls in the balance. To me, this is a perfect

FIG. 5 Vermeer, *Woman holding a balance.* (Reproduced with permission)

metaphor of our efforts to reach, through careful study and observation, the innermost reality, which is and will remain invisible.

And now that we have made the importance of the abstract clear, let us embark on our journey into the heart of theoretical physics, beginning with the concept of *limit*—the profound insight that escaped the Greeks but that made theoretical science possible.

3

Limits

It may be objected that there is no such thing
as an ultimate proportion of vanishing quantities,
inasmuch as before vanishing the proportion is not ultimate,
and after vanishing it does not exist at all ...
But the answer is easy ...
the ultimate ratio of vanishing quantities
is to be understood not as the ratio of quantities
before they vanish or after they have vanished,
but the ratio with which they vanish.

Newton, *Principia, Book I*, p. 442

Discontinuity from continuity

Unless you have received some serious mathematical training, the word 'limit' will not strike you as a mathematical term. It is a word of everyday language, in fact quite a poetic word. By acknowledging the existence of a limit you acknowledge an infinite world of possibilities beyond it. The Argentinian poet Jorge Luis Borges includes in his Personal Anthology no less than two poems with this title—both dealing with the finiteness of human experience. In one he laments 'a line by Verlaine that I will not remember', 'a threshold that I have crossed for the last time', 'a book that I will never read', and a 'mirror that waits for me in vain'.

Although I had not read Borges at the time, I was nevertheless puzzled when I saw 'Limits' as the title of a chapter in a book on mathematical analysis. What could this possibly have to do with the calculations which, I thought, were the essence of that discipline? I read the chapter avidly, but I was disappointed in the end. The idea turned out to be a rather obvious one, formulated with pedantic precision in terms of two small quantities denoted by the Greek letters ϵ and δ, of which the second could be chosen in such a way that a certain uninspiring requirement would be fulfilled for any value of the first. What a dreary definition! I remember college students trying to memorize the insidious formula while lazily lying on the green lawn of Miracles' Square in Pisa the day before the exam. To me this was the *non plus ultra* of formality—an exceedingly complicated way of saying that an infinite sequence of numbers can tend to a limiting value, just as the sequence of fractions 1/2, 2/3, 3/4, 4/5 ... approaches the value 1 more and more closely the further one goes.

But, I was wrong. Far from being a device for completing predictable lists of numbers, 'going to the limit' or 'taking the limit' is the trick that scientists and artists alike use to create new concepts and new spaces starting from the ones with which we are familiar. Actually, this became clear to me only after several years of work in theoretical physics. The point is that at the end of a limiting process you may find yourself in an altogether different world from the one in which you started. The landscape of this world depends sharply on the nature of the limit that has been taken or, when more than one limit is involved, on the *order* in which the limits have been taken. Even though the process of going to the limit has all the appearance of smoothness, its final product can be an object of a kind that did not exist in the sequence that led to it. For example, the sequence of fractions 1/1, 3/2, 7/5, 17/12 ...[1] tends to the square root of 2, which happens to be the ratio of the diagonal of a square to its side. This quantity is not expressible as a fraction—a discovery that sank Pythagoras and his followers in the depths of despair, for they believed that every meaningful quantity should be expressible as a ratio of integers. It is as if you had been walking across a shallow creek, being careful to put your feet on the emerging stones— and yet at the end of the journey you find yourself standing on water, or rather on a new kind of stone whose existence you had not appreciated until then. This is what I mean when I say that limits generate discontinuity from continuity.

The consequences of going to the limit are even more dramatic in physics than in mathematics. For, in mathematics, the new stones emerge as pure abstractions; in physics, they are statements about the behaviour of the natural world. The laws of physics are never laws about the world as it is, but about the world in a certain limit, or under a certain idealization. And for this reason, the relevant laws may change as we consider the world under different idealizations. In this chapter I will illustrate these ideas with simple examples taken from the most ancient branch of physics: mechanics—the science of motion.

Zeno's paradox

In a famous passage of his *Lectures on Physics*, Richard Feynman tells the following story about a woman driver and a police officer.[2] The officer stops the woman for speeding. 'I'm sorry ma'am I must give you a ticket:

---

[1] The rule is that the denominator of each fraction is the sum of the numerator and the denominator of the previous fraction; and the numerator of each fraction is the denominator of the same plus the denominator of the previous fraction.

[2] The gender bias of this anecdote earned Feynman the appellation 'sexist pig' from women's rights advocates. Later, when in truth it was too late, Feynman cleverly pointed out that the officer of the anecdote was also a woman.

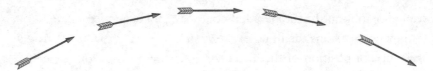

FIG. 6 Zeno's paradox of the arrow: what do we talk about when we talk about motion?

you were doing 60 miles an hour.' 'What do you mean?'—answers the woman—'I have not been driving for an hour!'

What should one reply to this? One might say: well, if you had continued to drive at this speed for an hour, you would have covered 60 miles. To which she could reply: 'But I would have never done such a thing! There are children crossing the road. No, I swear I would have never done this!'

The woman's objection is not devoid of merit, for our argument is somewhat circular. We are saying that the law forbids driving in such a way that, if maintained steadily for an hour, it would make us cover a distance of 60 miles. But we could also say: in such a way that, if maintained for a second, would make us cover a distance of one sixtieth of a mile. The second formulation makes it more difficult for the lady to claim good intentions, but still does not explain what 'that way' is, which the law does not allow, not even for an infinitesimal fraction of a second. 'That way' is, of course, 'a speed of 60 miles an hour'—but that is precisely what we wanted to define in the first place.

This difficulty must necessarily arise whenever one tries to explain velocity in terms of ordinary space and time intervals—for it cannot be done. Velocity does not live in ordinary space, and is not just the ratio of a distance to a time. Rather, it lives in an abstract space, which is tangential to ordinary space and is created by a process of limit. Indeed, velocity provides us with a basic example of discontinuity from continuity—a new concept arising from the limit of pre-existing ones.

I believe that the transcendental nature of velocity was first noticed by the Greek philosopher Zeno, and used by him to challenge the reality of motion. In the paradox of the arrow, which is named after him, Zeno envisions the flight of an arrow as a sequence of still frames showing the position of the arrow at successive times (Fig. 6). Photography had not yet been invented, but this is precisely what the photographer Eadweard Muybridge would do, many centuries later, to 'freeze' the race of a galloping horse. In each frame the arrow appears to be at rest, peacefully filling its own space; there is no hint of motion. So, asks Zeno, how do we know that the arrow is moving? If motion were real, then surely there should be a way to see it, to observe its presence in each frame. But this is not the case. On the contrary, the better our picture is (in modern language we would say, the shorter the exposure time of a frame is) the less evidence of motion we see. And if there is no motion in each frame, how can there be any motion at all?

To appreciate the depth of Zeno's concern, think of what you see when you watch a movie. You know that only one frame is present on the screen at any given time, then another, then another: nothing moves, but everything conspires to produce the illusion of motion. Well, I don't know if this is exactly what Zeno had in mind when he denounced motion as an illusion, but I can assure you that even today there are those who believe that the universe is a digital screen whose pixels change in colour and brightness according to pre-established rules, creating an impression of motion where there is only computation.

Be that as it may, Zeno's paradox has a natural resolution if we are willing to expand slightly our view of reality. Ordinary space should be supplemented with another space (invisible to the eye) whose points represent 'instantaneous velocities'. This space is created by a limiting process, which I will describe momentarily. The arrow will then occupy a position both in ordinary space and in 'velocity space', and the change of position in ordinary space will be controlled by the 'position' in velocity space.

Here is how you enter 'velocity space'. The idea is to compute an average velocity (not to be confused with true velocity) over shorter and shorter periods of time. An average velocity is obtained by dividing the distance covered during a certain time interval by the duration of that interval. Both quantities are finite, and their ratio is calculated by the ordinary rules of division. However, it is only when the time interval tends to zero that true velocity emerges as a sharp concept. It is the limit of the average velocity (an ordinary ratio of distance and time) when the duration of the time interval tends to zero. As Newton explains in the passage quoted at the beginning of this chapter, this limit is not the ratio of distance and time before they vanish or after they have vanished, but the ratio with which they vanish. It is the *rate of change* of position with time.

Figure 7 helps us to understand the concept in geometrical terms. The motion is described by a continuous curve which shows the position x of a car, a golf ball, whatever . . . as a function of time. To calculate the average velocity for the time interval between t_1 and t_2 we mark on the curve the two points that correspond to t_1 and t_2 and join them with a straight line. The average velocity is directly proportional to the slope of this straight line: the steeper the slope, the higher the average velocity. As we reduce the size of the time interval we see that the slope of the line varies, but ever more slightly. In the limit of vanishing time interval the straight line touches the curve at only one point. Such a straight line is called the *tangent* to the curve, and it is precisely the slope of this tangent that defines the instantaneous velocity at that time: the steeper the tangent, the higher the instantaneous velocity.

Looking back at Fig. 7, we see that our limiting procedure has replaced the actual motion of the car, which is represented by the smooth curve, by a straight line—the tangent—which touches the curve of the actual motion at just one point. The straight line describes a possible motion in its own right, a motion that could have taken place, theoretically, if the car had been travelling with a constant speed for all the past and were to keep the same speed for all the future. We have thus created a tangential reality, in which the car is damned to persevere eternally in its present state of motion. Needless to say, this tangential

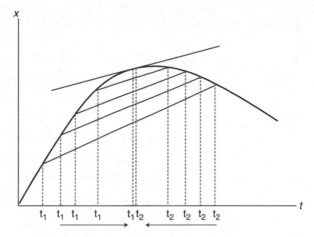

FIG. 7 Velocity as a limit. As the length of the time interval $t_1 - t_2$ tends to zero, the slope of the line that connects the corresponding points on the graph of position versus time tends to a limiting value. In the limit in which t_1 and t_2 coincide, this line becomes the tangent to the curve.

reality is purely conjectural—it did not happen and could not happen in real life, but its value lies not in having or not having happened; it lies in giving a sharp meaning to a concept—the concept of instantaneous velocity.

The abstract nature of velocity becomes even more evident when we consider a motion that is restricted to a line or to a surface in three-dimensional space. For example, Fig. 8 shows a bead that is bound to move along a curved wire. At each point, the velocity of the bead can have any magnitude, but its direction must always line up with the tangent to the wire, pointing either 'forwards' or 'backwards'. Notice that each point on the wire has its own private tangent, which is completely different from the tangent at a neighbouring point, no matter how near that point is. The jagged-looking ensemble of all the tangents constitutes a 'tangent bundle' grafted on top of the wire (Fig. 9). So while the bead lives, so to speak, on the wire, its velocity lives in a completely different and abstract space: the tangent bundle to the wire.

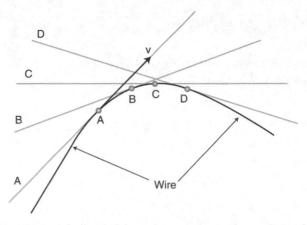

FIG. 8 The velocity of a bead sliding along a wire is always directed along the tangent to the wire. Each point of the wire has its own tangent.

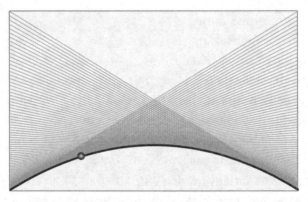

FIG. 9 The bead is constrained to slide along a curve, but its velocity lives in a more complex space—the 'tangent bundle'—which is the union of all the tangents to the curve.

The situation I have just described is not at all unusual. Scientific theories, like works of art, live in tangential realities that are conjured up by a limiting process. Starting from familiar concepts we dive into a fantastic space, navigate it according to certain rules, and re-emerge on the level of reality with a new concept, a new figure of thought— velocity in this case.

Two degrees of separation

One of the most beautiful things about abstract concepts is that they grow on each other like crystals of salt on a twig in a salt mine. You throw a leafless bough into the salt pit. Two or three months later, you find it covered with a shining deposit of crystals. The tiniest twigs are encrusted with an infinity of little crystals. The original branch is no longer recognizable.[3]

No sooner have we finished defining 'velocity' than we are irresistibly tempted to a higher level of abstraction. Like position, velocity can change in time, producing a motion in what we have called 'velocity space'. To this higher order motion we can apply the same concepts that served us well in analysing the motion in ordinary space. We can define an average rate of change of the velocity by dividing the change in velocity by the time during which it occurs. We go to the limit of vanishing time interval and by so doing we create a new concept, known as 'acceleration'. Acceleration is the rate of change of velocity in time, just as velocity is the rate of change of position in time. It is velocity in velocity space.

Acceleration (together with its inseparable companion—force) is the main actor on the stage of Newtonian mechanics—the quantitative science of motion that was initiated by Galileo in the seventeenth century. The non-intuitive character of the acceleration concept—that is to say, the *two* levels of abstraction that separate it from visible reality—is probably the main reason why mechanics took so much time and effort in the making. Because our senses are naturally attuned to following the trajectories of moving bodies, velocity is easily pictured as a vector pointing along the direction of motion. The doubts one can entertain about velocity are of such a philosophical nature that only an unusually critical or overactive mind like Zeno's will conceive them. Tradition has it that when Zeno challenged the reality of motion,

---

[3] Stendhal, *On Love—The Salzburg bough.*

Diogenes of Sinope came forward as an opponent—he did not say a word but merely paced back and forth a few times, thereby assuming that he had sufficiently refuted him. On the other hand, acceleration is difficult to visualize, even for physicists. Like velocity, it is a vector, namely a quantity endowed with a definite direction. The trouble with acceleration is that an object can move in one direction and at the same time accelerate in a different one. So, at variance with everyday language, the word 'acceleration' in physics does not necessarily refer to an increasing speed, but describes any deviation from uniform rectilinear motion, any change of velocity, whether it happens by increasing or decreasing the speed or simply by changing the direction of motion.

The realization that velocity and acceleration are independent quantities lies at the heart of Newtonian mechanics. I will be referring to this

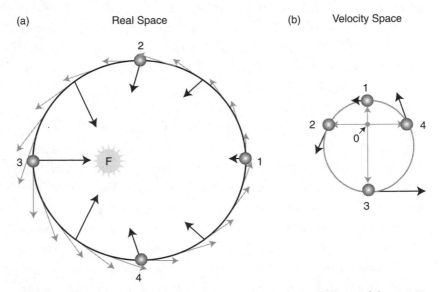

FIG. 10 (a) The velocity and the acceleration of a planet in orbit around the sun are shown as light and dark arrows respectively. Notice that the velocity is always tangent to the orbit, and the acceleration is always directed towards the sun. (b) The orbit of the same planet in *velocity space* is a circle. Again, light arrows and dark arrows indicate velocity and acceleration respectively. Just as the velocity is tangent to the orbit in real space, the acceleration is tangent to the orbit in velocity space. The sun is invisible in this graph. 'o' denotes the point of zero velocity.

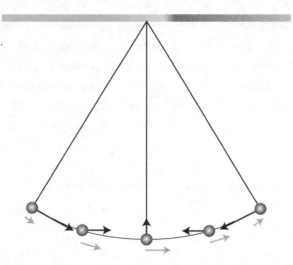

FIG. 11 The velocity (light arrow) and the acceleration (dark arrow) of a pendulum at various positions during a swing.

key feature as to the *acceleration–velocity split*. For example, a planet going around the sun has a velocity tangential to the trajectory, but its acceleration points steadily towards the sun, as shown in Fig. 10. This figure shows the motion of the planet first in real space (a), then in velocity space (b) (see caption for details). As for the simple pendulum bob shown in Fig. 11, observe how its acceleration keeps swinging back and forth between horizontal and vertical directions—a completely non-intuitive behaviour that can draw gasps of surprise even from an audience of stone-hearted physicists! The difficulty in understanding these figures is that we don't have an organ to sense motion in velocity space, except that earnest but fragile one—our mind.

Galileo and Aristotle

In the late sixteenth century Galileo was perhaps the first to hit on the right idealization on which the modern science of motion was to be built. He understood that a free body moving in space in the absence of forces would keep going forever in the same direction and with the same speed, without any acceleration. But he didn't state the idea as a formal Law. In the year of Galileo's death (1642) another genius was born who elevated that simple-sounding idea to the status of *Principle of Inertia*, or the *First Law of Dynamics*, and went on to build upon it the edifice of classical dynamics. In Newton's own words:

> Every body perseveres in its state of being at rest or of moving uni-formly straight forward, except insofar as it is compelled to change its state by forces impressed.

For those who imagine science to be based on hard evidence the Principle of Inertia is a remarkably vague statement, almost a metaphys-ical one. A body persevering in a state of uniform straight forward motion: who has ever seen such a thing? How to make it happen? How to live long enough, or travel far enough to ascertain that it is indeed so?[4] The one case in which the statement seems easy to verify is that of an object which is initially at rest. Then the principle says that it will remain at rest forever, unless compelled by a force to do otherwise. Aristotle would have readily agreed to this, for he believed that motion cannot occur without a mover. But the principle of inertia goes much further, saying that there is no qualitative difference between rest and motion with constant velocity, that rest is a limiting case of motion, that motion is a different representation of rest.

Aristotle's view that motion requires 'force' was supported by com-mon sense and observation. Nowhere is this more evident than in the biological world, for life is a continuous struggle to preserve motion against the immobility of death. Here is, for example, a passage from Jack London's novel *White Fang*—a book that has a lot to say about the Laws of Nature:

---

[4] The very concept of an infinite straight line is quite problematic. The possibility of extending a finite segment of straight line into an infinite line is a postulate of geometry, not something that can be proved. And drawing an absolutely straight line is a highly non-trivial matter if you don't already have one available (i.e. a ruler).

It is not the way of the Wild to like movement. Life is an offence to it, for Life is movement; and the Wild aims always to destroy movement. It freezes the water to prevent it running to the sea; it drives the sap out of the trees till they are frozen to their mighty hearts; and most ferociously and terribly of all does the Wild harry and crush into submission man— man who is the most restless of life, ever in revolt against the dictum that all movement must in the end come to the cessation of movement.

In comparison to Aristotle's, Galileo's view of motion must have seemed far-fetched and fantastical. If that view was right, then motion on a straight line was the natural state of affairs and no external intervention, human or divine, was needed to maintain it. The problem was not to explain the rectilinear motion, but the deviations from it.

Galileo offered many arguments in support of the principle of inertia. Here is a particularly elegant one, which nicely illustrates the power of limits. Consider the experimental set-up shown in Fig. 12, in which the surfaces of the two ramps and the connections between them are all made very smooth by careful polishing. Then a smooth object (say an

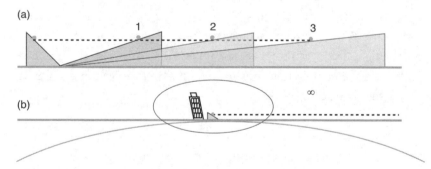

FIG. 12 Thought experiment leading to the principle of inertia. In the experiments of part (a) an object descends along the left slide with increasing speed and rises along the right slide with decreasing speed until the initial height is recovered. Replacing slide 1 with slide 2, and then with slide 3, the distance travelled by the object increases. In part (b) we go to the limit in which the right slide is flat. Now the initial height can never be recovered, and the object has no choice but to continue to travel on a horizontal line without acceleration or deceleration. The oval encloses the part of the ideal set-up (including the leaning tower of Pisa and the surface of the Earth) that is practically accessible to observation.

ice cube) released from the top of the left ramp will slide down and rise up the ramp on the right, up to exactly the same height from which it was released. This is a fact that *can* be verified experimentally. The smoother your surfaces are, the better you will find that the object recovers the initial height. Naturally, the speed of the ice cube depends on its height above the plane: it increases as the cube descends, it decreases as it climbs up the right ramp, and vanishes when it reaches the highest point.

Now imagine reducing the inclination of the right ramp. In order to recover the initial height the object must travel a longer distance. It is now even more important that the surface of the ramp be very smooth, and you may need to go through a second and third round of polishing to achieve the desired accuracy. But you can do it, and so you can verify that, as the inclination of the ramp decreases, the object continues to rise to the same height, even though this takes a much larger displacement in the horizontal direction.

Now comes the magic leap by which we go beyond the empirical observation and grasp the law. We go to the limit in which the inclination of the right ramp tends to zero, i.e. the right ramp becomes horizontal. What happens is that the body strives to go back to the initial height, but it cannot, because the ramp does not rise. What can the body do? The only option is to keep moving forever, and since its height doesn't change, neither does its speed. This is precisely what the principle of inertia says.

Quite ingenious, isn't it? Except that the last step is a leap in the abstract, into a literally tangential world in which neither the curvature of the Earth, nor the inevitable friction, nor extraneous forces, nor the finiteness of the lab, nor any other trivial contingency matters. All the steps that lead to the final leap belong to the real world, and must be executed with care and method. As you decrease the inclination of the right ramp, you must make sure to improve the smoothness of the surface, so that the ice cube still reaches the same height. If you keep decreasing the inclination of the ramp, without improving the smoothness of its surface, the experiment will eventually fail due to surface roughness. It is important that you keep reducing the angle and the roughness in the proper order. But, the experiment can only take you so far and no further. The final leap, the taking of the limit, is a mental process, and it is fantastic.

The crucial passage of the *Dialogue on the Two Systems of the World*[5] in which Salviati (speaking for Galileo) leads Simplicio (the honest but simple-minded Aristotelian philosopher) to concede the validity of the principle of inertia makes this point very clear. At first, Salviati seems to appeal to experimental evidence in support of his position. The issue is whether a stone dropped from the top of the mast of a ship, when the ship is moving, will strike the deck at the foot of the mast, or as far from it as the ship has advanced during the time of the fall. Simplicio, following Aristotle, believes the second scenario to be correct. This would spell trouble for the theory of the moving Earth, suggesting that a stone dropped from the top of a tower (the mast) should strike the ground far from the base of the tower (the Earth having moved during the time of the fall), in glaring contradiction with experience.

Now Salviati, *sublime intellect, who nourished himself only of exquisite speculations*, asks Simplicio whether he has ever made this experiment of the ship. Simplicio candidly admits he hasn't, then asks the natural question: have you? At this point one would expect Salviati to bury his opponent under an avalanche of experimental facts, but he doesn't. He himself has not done the experiment! Yet he *knows* what the outcome of the experiment will be, and proceeds to convince Simplicio with a beautiful and purely theoretical argument.

> Without experiment—Salviati says—I am sure that the effect will happen as I tell you, because it must happen that way; and I might add that you yourself also know that it cannot happen otherwise, no matter how you may pretend not to know it.[6]

[5] Galileo Galilei, *Dialogue on the Two Systems of the World*, Second Day, 171.
[6] Galileo Galilei, *ibid.*

The argument is even simpler than the one I outlined earlier. An object that is moving *up* on a polished ramp slows down until it comes to rest. Similarly, an object that is moving *down* on the ramp picks up speed as it descends. But an object that is moving neither up nor down, i.e. on a plane, is just on the vanishing borderline between these two possibilities, and must behave compatibly with both. Such an object can neither speed up nor slow down. Therefore its speed must remain constant. Now, when a stone is dropped from the mast of a ship it continues to move along with the ship with a constant horizontal velocity, as if it were on a horizontal plane, and therefore will always fall at the foot of the mast, whether the ship is moving or not.

Privileged frames

Convincing as they are, Salviati's arguments leave a troublesome point to be dealt with. The infinite plane upon which the object would be seen moving forever in uniform rectilinear motion must itself be free of acceleration in an absolute sense: otherwise the Principle of Inertia will not hold true. So much for our Universal Principle!

Imagine, for example, that the ideal plane of Fig. 12 is firmly anchored to the surface of the Earth at the North pole, so that it turns around completely in a period of 24 hours. You don't have to be a physicist to see that, *relative to such a rotating plane*, the trajectory of the object will not be a straight line but a curve. The point is that according to the Principle of Inertia the object must move on a straight line with respect to 'absolute space', whatever that means—perhaps the distant stars of the Universe. But precisely for that reason the same object will not move on a straight line with respect to our rotating plane.

So the supporters of Aristotle were not completely wrong in supposing that the rotation of the Earth should somehow influence the motion of things and creatures on the Earth. Their error was more subtle. They thought that the Earth's velocity would directly manifest itself, while Galileo and Newton realized that the Earth's velocity would remain invisible, and that any observable manifestation of the motion of the Earth must arise from the Earth's acceleration—a much weaker effect. Indeed, a very careful observation would have revealed that the trajectory of a falling body is not exactly vertical, but does deviate ever so slightly to the East because of the daily rotation of the Earth, proving that Simplicio was not entirely wrong and Salviati not entirely right in a narrow factual sense. This is a classic case in which seeing too much might have fatally hindered the progress of physics. It was wise of Salviati to gloss over that detail.

The principle of inertia is much more than a physical law: it is a declaration of faith in the existence of an absolute, infinite, and uniform space—the ideal arena in which free bodies persevere in a state of uniform motion. In fact, we now know that the principle of inertia is the prototype of what physicists call a *conservation law*: that is to say, the statement that a certain quantity remains constant in time. In this case, the conserved quantity is called 'momentum'. In Chapter 8 we will see that conservation laws reflect fundamental symmetries of the world. Then the Principle of Inertia, alias conservation of the momentum of a free particle, will be shown to be a consequence of the uniformity of space—the fact that all points in space are absolutely equivalent and no point is special.

Reference frames in which the principle of inertia appears to be violated do exist (one, as we saw, is the rotating Earth), but they are not to be trusted because they are themselves accelerated with respect to absolute space. The motion of bodies becomes transparent only when referred to an abstract reference frame which is not accelerated with respect to absolute space: it is only in such a privileged reference frame that the Principle of Inertia holds true. In physics, such a reference frame is called—with splendid circularity of thought—*an inertial reference frame*. And the principle of inertia simply asserts that such a reference frame exists.[7]

You may feel that the restriction to inertial frames is unsatisfactory and unfair. Why should we limit ourselves to the use of inertial reference frames when most reference frames, including the most important one, on which we live and die, are non-inertial? In fact, it turns out that the restriction is unnecessary. At the end of Chapter 6 we'll see that the laws of physics can be formulated in *any* reference frame, provided we are careful to make the necessary adjustments. But all these fancy developments do not really eliminate the idea of absolute space; rather, they teach us how to look at absolute space from different points of view, and how to connect the observations made from different points of view.

The laws of dynamics

As with every work of art, Newtonian mechanics creates its own characters. We have already encountered velocity, acceleration, and absolute space. The next character is force. Force is implicitly defined by the Principle of Inertia as the agent of accelerations. So if you observe a planet, say Mars, going around the sun in a curved path, you can be sure that there is a force acting on it, even if you don't know whence and why. And this was Newton's stance when he postulated the existence of a gravitational force, even though he could not explain its origin.

Once it is accepted that forces are the cause of accelerations the next natural question is: what is the relationship between acceleration and force? The simplest hypothesis is that the *net force*[8] needed to impart a certain acceleration is directly proportional to the acceleration, i.e. it doubles if the acceleration doubles, it triples if the acceleration triples

[7] Notice that if you accept the existence of *one* inertial reference frame, then you must accept the existence of infinitely many, because all the frames that move with constant velocity with respect to an inertial frame are also inertial.

[8] By *net force* we mean the sum of all the forces acting on the object, regardless of their origin.

and so on. However, this cannot be the whole story, because the acceleration imparted by a force must also depend on the quantity of matter that is being accelerated. The more massive the object, the larger the force that is needed to impart a certain acceleration. We call *mass*, and denote by the symbol *m*, the 'quantity of matter' in a body. Then we can say that the net force needed to impart an acceleration is proportional not only to the acceleration, but also to the mass, i.e. it doubles if the mass doubles and so on. This is the content of Newton's *Second Law of Dynamics*,

$$\vec{F} = m\vec{a}$$

where the arrows above the symbols for the force and the acceleration remind us of the fact that these quantities are vectors—that is to say, quantities with a definite direction.

Of course, I have been talking of mass as if we already knew what it is, which is not quite the case at this point. Mass, it turns out, is the last protagonist of Newtonian mechanics—a character born out of the equation of motion $\vec{F} = m\vec{a}$. But, wait a moment: how can the equation for the force, $\vec{F} = m\vec{a}$, tell us what mass is, if mass is needed in the first place to tell us what force is? Clearly, some additional input is needed to escape this logical loop.

The input is Newton's *Third Law of Dynamics*. When two bodies A and B interact, for example, in a collision, they exert forces on each other. According to Newton's Third Law the force that A exerts on B during the collision is equal in magnitude and opposite in direction to the force that B exerts on A.

In essence, this is a way of saying that force is something that is *exchanged* between two bodies: each body pushes and is pushed at the same time. This is inevitable. When you kick a football some abstract thing, 'momentum', flows from your foot into the ball. The ball that receives the momentum is pushed; the foot that delivers the momentum is pushed too, but in the opposite direction. An isolated body cannot exert a force on itself.

With the help of the Third Law we can define precisely what mass is. We shoot the two bodies against each other and observe their accelerations. Since, according to Newton's Third Law, the same magnitude of force acts on each body, we can immediately conclude that the two bodies will accelerate in inverse proportion to their masses. If, for example, one body has twice the mass of the other, then its acceleration will be half the acceleration of the other. The ratio of the masses determines the ratio of the accelerations, and, conversely, the ratio of the accelerations determines the ratio of the masses.

At this point all we need is a standard body to which we assign, by convention, a mass of 1 kilogram. Then the mass of another body can be determined, in principle, by shooting it against the standard 1 kilogram body and observing how its acceleration compares with the acceleration of the standard body. This is how we are able to define masses from accelerations without knowing anything specific about the forces, except the general fact that they occur, inevitably, in pairs.

Illuminations

Newtonian mechanics is not an intuitive theory. You don't realize this until you try to apply it to the solution of concrete problems. You may feel that you have perfectly understood the content of the Laws, and you may be a firm believer in the Principle of Inertia—but working out problems will be a severe test of your faith. You'll discover inside you a Peter ready to betray Jesus at the first challenge. Over and over you will catch yourself slipping into the more natural non-inertial (Aristotelian) way of thinking, or you will fail to appreciate the import of the acceleration–velocity split, or both. In the next two figures I illustrate two aspects of Newtonian subtlety.

Figure 13 shows the motion of a body which is subjected to an alternating force: a force that pushes for one second to the right, then for one second—and with the same intensity—to the left, then again to the right, and so on. According to Aristotle, or to naive common sense,

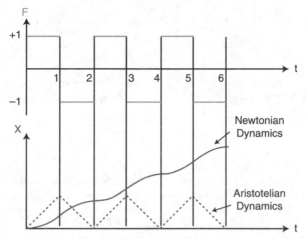

FIG. 13 According to 'Aristotelian Dynamics' (velocity proportional to force), a body subjected to an alternating force F (plotted in the top panel) would move back and forth, returning periodically to the starting point, as shown by the dashed line. The correct behaviour, predicted by Newtonian Dynamics (acceleration proportional to force) is shown by the solid graph. The body keeps advancing, with variable velocity, in the direction of the first push.

the body should simply move back and forth, returning to the starting point every two seconds. But this is completely wrong! According to Newton's Laws, the body will keep moving indefinitely in the direction of the first push. What happens is that the first push accelerates the body to the right, and the subsequent pull brings the velocity back to zero, *but does not restore the original position*. In fact, the velocity is always directed to the right, so at the end of the first push-pull cycle the body will have moved some distance to the right. The second, third,... sixth cycle are just repeats of the first, so the body keeps moving steadily to the right. This is a rough, but essentially correct model of walking: you push yourself forwards with your back foot,[9] then you slow yourself down with the front foot, and so on and so forth. Or, if you want a fancier

---

[9] More accurately, I should say that you push with your foot against the ground, causing the ground to push back against your foot, thus propelling you forwards.

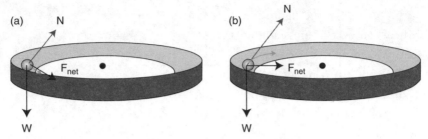

FIG. 14 Two illustrations of Newtonian mechanics. Observe the subtle difference between (a) ball rolling *down* the track and (b) ball rolling *around* the track. The difference between the two situations is in the force N exerted by the track on the ball. This force is slightly larger in case (b) than in case (a). In case (a) the force N neutralizes only the component of the force of gravity perpendicular to the track. The parallel component survives and pulls the ball down the track. In case (b) the force N completely neutralizes the force of gravity, producing in addition a horizontal force which points towards the centre of the circular track and keeps the ball rolling on a circular path.

analogy, think of the way governments operate under conditions of checks and balances. In order to forward its agenda the executive breaks the law (the push); a little later it is stopped by an order by the Supreme Court (the pull); but, in the interval between the push and the pull, the government has managed to forward its illegal agenda (the prisoners have been tortured).

Figure 14 shows a classic example of circular motion, which can be found in any textbook of elementary physics. Most of us are prepared to accept that two bodies, subjected to identical forces, might evolve in completely different ways if they start from different initial conditions. But in this example different initial conditions produce different forces, which in turn lead to completely different motions.

In part (a) of the figure the ball, initially at rest, rolls down a circular track which forms an angle with the horizontal. In this case the component of the force of gravity perpendicular to the track is neutralized by an opposite force exerted by the track on the ball. The remaining component of gravity, parallel to the track, causes the ball to roll down.

In part (b) of the figure the same ball is imparted an initial velocity, which causes it to press a little harder against the track. Now the vertical component of the force exerted by the track on the ball neutralizes completely the force of gravity, so the ball does not roll down. At the same time, there is a residual horizontal component of the force exerted by the track on the ball, which pushes the ball towards the centre of the track. This force is similar to the force with which the sun attracts a planet in its orbit—although its origin is completely different! Just like a planet going around the sun, the ball rolls around the track in a horizontal circular path.

Newton and Kepler

Among many spectacular applications of Newtonian mechanics, such as the explanation of the tides, the prediction of the existence of planet Neptune, the accurate calculation of the orbit of a space probe—all flowing from the concerted and delicately balanced action of the three basic laws—none is so impressive as the deduction of the universal law of gravity.

Gravity is the force that holds together the solar system, powers the stars, sweeps the oceans with majestic tides, and yet does not disdain pulling little apples to the ground. According to theory, any two pieces of matter anywhere in the Universe attract each other with a force that is inversely proportional to the square of the distance between them, and directly proportional to their masses. This law—in particular, the pro-portionality of the force to the inverse of the square of the distance—was discovered by Newton, with essential input from Johannes Kepler.

Kepler had observed that the orbits of the planets around the sun are not circles, as Copernicus and Galileo had believed, but ellipses. The precise statement of *Kepler's First Law* is that the orbit of a planet is an ellipse with the sun at one of the foci (Fig. 15).[10] In the course of the

---

[10] Think of the ellipse as a circle stretched in one direction, or as the set of all the points of the plane whose distances from two special points, known as the foci, add up to a given constant value.

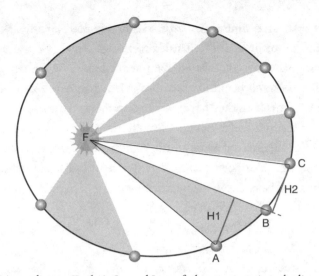

FIG. 15 According to Kepler's Second Law of planetary motion, the line that joins the sun to a planet sweeps equal areas in equal times. In this figure, the area of an elliptical orbit has been sliced into ten equal sectors, and each sector is swept in one tenth of the total time of revolution. In the end, we will go to the limit in which the number of sectors tends to infinity, while the width of each sector tends to zero. The areas of the adjacent triangles FAB and FBC are given by the products $\frac{FB \times H_1}{2}$ and $\frac{FB \times H_2}{2}$ respectively, where H_1 and H_2 are the heights of the two triangles, relative to the common base FB (recall that the area of a triangle is the base times the height divided by 2). The error one makes in replacing the curved sides of the triangles by straight lines becomes negligible in the limit in which their length tends to zero. The equality of the two adjacent areas implies $H_1 = H_2$. Now observe that H_1 and H_2 are components of the displacement in the direction *perpendicular* to the radius FB. The fact that these displacements occur in equal times, implies that the component of the velocity perpendicular to the radius has the same value in two adjacent sectors. This, in turn implies that there is no component of the acceleration perpendicular to the radius, and hence no component of the force in that direction. We conclude that the force of gravity must be entirely directed *along* FB, i.e. towards the sun.

motion the distance of the planet from the sun varies. When the planet is closer to the sun it moves a little faster than when it is farther away from it. Kepler discovered that, in spite of this variation, the line that joins the sun to the planet sweeps equal areas in equal times. This fact is now known as *Kepler's Second Law*. Apparently, the changes in velocity and distance from the sun are so finely attuned that the area swept by this line in a given time (say one hour) is the same no matter where we are along the orbit (Fig. 15). It can be easily shown that this regularity implies that the force acting on the planet is directed towards the sun. (The proof is sketched in the caption of Fig. 15.) Notice that, according to Aristotelian mechanics, such a force would inevitably pull the planet towards the sun, where it would be destroyed. What saves the day is the fact that the force determines the direction of the acceleration, not that of the velocity, which can therefore continue to point along the orbit. This is the characteristic acceleration–velocity split of Newtonian mechanics, already noted in Fig. 14(b).

Kepler had also discovered that the *square* of the time it takes a planet to complete a revolution around the sun is directly proportional to the *cube* of the average distance of that planet from the sun (*Kepler's Third Law*). Thus, a hypothetical planet whose distance from the sun was four times the distance between Earth and sun would take eight years to go round the sun, because $8^2 = 4^3$. This simple observation is sufficient to establish that the force of gravity between two bodies must be proportional to the inverse of the square of their distance. Let us see why.

Let us compare the accelerations of two planets that travel around the sun on circular orbits, of which one has four times the radius of the other. You may immediately object that no such planets exist in the solar system or elsewhere: didn't we just say that the orbits of planets are ellipses? Yes, but our point is that such planets *might* exist, for a circle is, after all, a special kind of ellipse, obtained in the limit in which the two foci coincide in a single centre. And if Kepler's law and the law of mechanics are as universal as the law of gravity, then they must be prepared to give the right answer even in the unlikely eventuality of circular orbits. So let us see what the two laws predict in this case. According to Kepler's Third Law, the period of revolution of the outer planet must be eight times that of the inner planet. What about the acceleration? Recall that acceleration is distance/time/time. Since in going from the inner to the outer orbit distance has been magnified fourfold and time eightfold, we see that the acceleration of the outer orbit must be reduced by a factor $4/8/8 = 1/16 = 1/4^2$. Now back to the laws of mechanics. We know that the acceleration is proportional to the force. This means that the force of gravity on the outer planet is 4^2 times smaller than the force on the inner planet, while the distance is 4 times larger. The fact that the gravitational force on the outer planet is not 4 but 4^2 times weaker than on the inner planet proves that the gravitational force is inversely proportional to the *square* of the distance: which is what we wanted to show.

This and other results would have probably remained hidden from sight if people like Kepler, Galileo, and Newton, had not had the intellectual courage to abandon a point of view that was bolstered by common sense and by authority, go to the limit, and embrace a more abstract and, to some extent, fantastical view of the universe. In the case of Newton's theory of gravity this went so far as to contemplate the almost mystical notion of a force acting across empty space, without the mediation of matter. We know that Newton himself found this idea 'inconceivable' (as if he had not already conceived it!) and yet did not shrink away from it, nor tried to propose a more plausible alternative. Perhaps, like Feynman, he thought that it is better 'to live not knowing, than to have answers that might be wrong'.

Aristotle and Newton

One should not be so dazzled by the success of Newtonian mechanics as to forget the reasons for the partial success of Aristotelian mechanics—the mechanics that makes velocity, rather than acceleration, proportional to the force. How could a theory be so wrong in its premises and yet give such a reasonable and convincing description of Nature as to satisfy the world's best minds for several centuries? The theoretical physicist Alexander Polyakov has written that 'any non-trivial idea is in a certain sense correct',[11] and here we can see the truth of that assertion.

[11] A. M. Polyakov, *Gauge Fields and Strings*, p. 1.

Undoubtedly, it is precisely Newtonian mechanics that gives us the tools to understand why Aristotelian mechanics works so well in so many cases. This is why we think of Newtonian mechanics as a much deeper theory than Aristotelian mechanics, and insofar as it is deeper it is also closer to the truth. The details are quite complicated, but the basic idea is easy to grasp. Polyakov's 'certain sense' in this case is just a matter of limits.

Newtonian mechanics, being based on the Principle of Inertia, is, first and foremost, the mechanics of point particles moving in empty space. But what is a 'point particle'? As a student I was stunned when a girlfriend of mine asked me this question. The concept of point particle was so hard-wired in my brain that I had considered it absolutely self-explanatory until then. Actually the abstract concept of 'point particle' covers an incredible variety of things. Perhaps the image of the Democritean atom presents itself first, but we all know that even enormous objects like planets and stars can be treated as point particles, if certain conditions are met. Namely, the size of the object must be small in comparison to the size of the region of space in which it moves (for example, the size of a planet is much smaller than the size of its orbit). Furthermore, any internal motion of its parts must be either absent or irrelevant. Every material object, regardless of size, shape, and composition, becomes a point particle in the appropriate limit, i.e. when we look at it from such a large distance that it appears as a single point.

Now let us say that our 'particle' is, in fact, a parachute which is gently descending through air to deliver a skydiver to the safety of land (Fig. 16). We observe that the parachute descends with essentially constant speed and zero acceleration, except for a brief period of turmoil immediately after being deployed. This behaviour seems to be in good agreement with Aristotle's law of motion (speed proportional to force) and in glaring disagreement with the supposedly correct Newtonian mechanics (acceleration proportional to force).

But wait, Newtonian mechanics says that the acceleration is proportional to the *net* force, that is to say, the sum of all the forces acting on the body. Gravity is just one of these forces, but there are others. These

FIG. 16 The motion of a parachute is well described by a brand of 'Aristotelian Dynamics' in which a constant force produces a steady velocity. (Photo by Franz Frank, released to the public domain.)

forces arise because our 'particle'—the parachute with its load—is not alone in a vacuum but is moving through air, which implies the presence of more actors than meet the eye. The invisible actors are molecules of air, which continually strike the surface of our parachute from above and from below, but, on average, more frequently from below than from above, since the parachute is moving downwards. During each collision the molecule exerts a force on the parachute—a force that is directed upwards when the molecule hits the parachute from below. The actual time-structure of this force is completely chaotic and well beyond our power to calculate (for a rough analogy, think of raindrops randomly striking the roof of a cabin, only much more frequent, say

10^{22} molecular collisions per square centimetre of surface per second). But it is precisely the complexity of the problem that allows a simple approach to its solution. As a first step we must relinquish any hope of following individual molecular collisions: we shall be content with an average description of what is going on during time intervals that are very long compared to the typical interval of time between molecular collisions, and yet far too short for the parachute to have moved significantly. We can then show that the net result of all these collisions is to produce, on average, a drag force, which is directed upwards, and increases as the speed of the parachute increases. Taking the limit, in this case, means observing the system through a time window that is very large compared to the typical time interval between molecular collisions, but still very small compared to the time it takes the object to change its position significantly. Within this *time window* we have the miracle of a Newtonian system which appears to follow the laws of Aristotelian mechanics.

How is this possible? Like this. After a short initial period, during which the parachute slows down the fall of the skydiver, a steady regime is reached, in which the force of gravity is compensated (on average) by the drag force: the net force is then zero, and the parachute descends with constant speed. Aristotle and Newton predict the same behaviour, but for very different reasons. Aristotle would say that the pull of gravity causes the parachute to descend with a constant speed. Newton would respond that the very fact that the speed is constant means that there is no net force (on average) acting on the thing. To actually see the difference between the two theories you should refine your observation to the point that you are able to notice tiny accelerations due to fluctuations in the rate of molecular collisions (like raindrops hitting the roof of our cabin, air molecules hit the parachute sometimes harder, sometimes less hard than average). Only at this extremely high level of accuracy would you be able to verify that Newton is right—that the speed of the parachute is not really constant, but rapidly fluctuating about a constant value, and that the acceleration is proportional, at each time, to the instantaneous value of the net force. But, on such a fine scale, the very concept of drag force disintegrates, replaced by the much more accurate concept of molecular collisions. So the 'drag force' emerges as a useful concept only when the microscopic details are glossed over or, as Musil would have it, when the dust of the 'inessential facts' is swept away.

Aristotle with a twist

It is important to appreciate that in the previous section we have not derived Aristotelian mechanics as a special case of Newtonian mechanics. The two theories are based on incompatible principles which prevent the former from fitting snugly inside the latter. What we have done is, in a sense, much more exciting, for we have shown that in a certain observational window the behaviour of a system that obeys Newton's laws of motion is perfectly well described by Aristotelian mechanics—a completely different theory. We see that the Aristotelian mechanics provides the *effective theory*, in a world in which Newtonian mechanics is the *fundamental theory*. You might wonder whether by suitably choosing the phenomenon and the window of observation, as we have just done in the case of the parachute, we might get Newtonian mechanics to behave in ways that are neither Aristotelian nor Newtonian in the conventional sense. Could we perhaps generate a third type of effective dynamics, starting from particles that obey Newton's law of motion?

Absolutely! And here is an example. Consider this familiar toy, a top, which is rapidly spinning about its axis as shown in Fig. 17. It is important for our argument that the spinning motion be very fast, much faster than any change in the orientation of the axis of the top. In fact, you may have already guessed that we are going to consider the limit in which the spinning frequency tends to infinity. In this limit a remarkably simple and utterly counterintuitive dynamics emerges from Newton's laws. You can verify it for yourself: the spinning top, once started in a nearly vertical orientation, does not drop to the floor, but

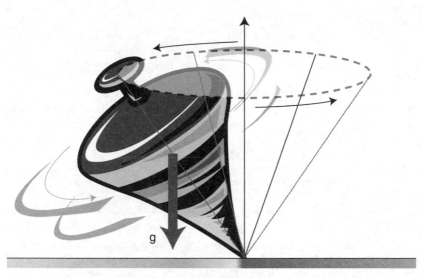

FIG. 17 A rapidly spinning top does not fall under the action of gravity. The force of gravity—indicated by the thick vertical arrow—drives a slow rotation of the axis of the top around the vertical.

starts turning about the vertical, the tip of the axis describing a circular motion in a horizontal plane. This motion is called 'precession' for astronomical reasons: as a motion of the Earth's axis, it causes the equinox to arrive every year a little earlier than the previous year, i.e. to *precede* the appointment. What we have here is Aristotelian dynamics with a twist. The speed of the precessional motion (i.e. the rotation of the axis of the top about the vertical) is directly proportional to the force of gravity, as Aristotle would have it, but its direction is in the horizontal plane, at 90° to the direction of gravity. The top falls sideways, so to say.

It is quite a challenge to explain how this strange global behaviour emerges from the Newtonian dynamics of the point particles which make up the top. The key feature is the *rigidity* of the top—the fact that its constituent particles are locked in fixed relative positions. Considering that each particle is subjected not only to the force of gravity, but also to the forces exerted by the other particles, we see that the problem

(a) (b)

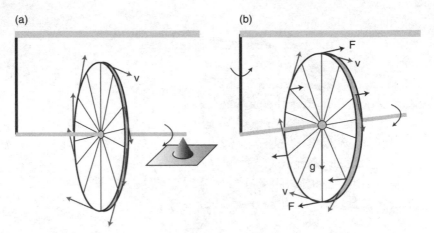

FIG. 18 'Falling sideways'. In part (b) of the figure we see how the force of gravity (g), pulling downwards on the hub of a rapidly spinning wheel is converted by the spokes into sideways forces (F) acting on the rim (dark arrows). These forces change the directions of the velocities of the particles along the rim (light arrows). The result is that the axis of the wheel turns in a horizontal plane.

is very complex, and that the emerging global motion might be very different from the intuitive behaviour, as indeed it is.

In Fig. 18 the 'sideways falling' of the top is analysed from a Newtonian point of view. For clarity's sake I have replaced the top by a spinning bicycle wheel, with one end of the axle attached to the ceiling, the other free. The free end of the axle is initially resting on a support, with the wheel spinning at full regime: this initial condition is depicted in part (a) of the figure. Now we remove the support and let the free end of the axle go. What happens? The force of gravity, no longer balanced by the support, pulls down on the axle and, through the spokes, exerts a force on the particles that constitute the rim of the wheel. This is the essential effect of rigidity: to convert a vertical force, the weight of the wheel, into a horizontal force exerted by the spokes on the rim of the wheel. This horizontal force, according to Newton's Second Law, produces an acceleration. In part (b) of the figure you see that the force acting on the particles at the top of the wheel is directed to the right, while the force acting on the particles at the bottom of the wheel is directed to the left. Because the wheel is spinning, the particles located at the top and at the bottom of the wheel are already moving fast in the direction of the vector **v** of Fig. 18. Therefore, the relatively weak force exerted by the spokes in the perpendicular direction **a** can only change the direction of this motion, deflecting the particles at the top of the wheel to the right, and those at the bottom of the wheel to the left, as shown in Fig. 18. The net result of these accelerations is to change the orientation of the wheel: the axle rotates in a horizontal plane rather than dropping vertically down. The relevant acceleration here is not the large one associated with the fast spinning motion of the wheel, but the small one associated with the global change in the orientation of the wheel, i.e. the speed of the precessional motion!

And now we see the origin of the apparently non-Newtonian behaviour of the precessional motion. The point is that the force of gravity accelerates the particles, but this acceleration manifests itself as speed of precession, so that in the end one can legitimately say that the speed of precession is directly proportional to the force that produces it.

Vortices

The motion of a top is more than a mere curiosity. In physics as in poetry you cannot talk about something without talking at the same time about everything else. The spinning top is a model of behaviour that shows up in many different and apparently unrelated contexts. For example, in the motion of fluids. Hydrodynamics (the science of the motion of fluids) does have its own spinning tops, which have been known since antiquity and have caught human imagination no less than 'point particles'. They are called 'vortices'.

No one has failed to observe the small vortex forming when a basin full of water empties through the drain. Far from the drain the water circles slowly, so slowly that one can hardly notice its motion. But as one approaches the centre of the vortex, which is located just above the drain, the circulation speeds up and the surface of the water bends downwards to form a nearly vertical edge, which sucks in everything that happens to be floating nearby.

In the seventeenth century, René Descartes suggested that planets might be dragged around the sun by a powerful vortex. Newton took the suggestion seriously, calculated its consequences, and found that it could not be so. The problem is that the speed of the fluid in a vortex is inversely proportional to the distance from the centre: so, if Descartes were right, we should find that the period of revolution of a planet is inversely proportional to the *square* of its distance from the sun. But this contradicts Kepler's observation that the square of the period is inversely proportional to the cube of the distance. You see that fantasy is only half of the story: you need precision too!

Fantastic literature is full of stories of unlucky ships devoured by vortices. One of the favourites of my childhood is *A descent into the Maelstrom*, by Edgar Allan Poe. The narrator, caught in a terrifying tidal vortex, manages to escape death by (i) giving up hope, (ii) making good use of what would have otherwise been the last panicky minutes of a wasted life by studying the motion of objects falling into the vortex, (iii) discovering that cylinder-shaped objects are sucked in more slowly than spherical ones, (iv) finding salvation in a cylinder-shaped cask.

In spite of its daunting complexity, a vortex can be regarded as a spinning point particle as long as we look at it on a scale that is much larger than the size of the inner core region (a limit again). The position of this 'particle' coincides with the centre of the vortex and the spin corresponds to the circulation of the fluid about the centre. This nice picture breaks down inside the core, just as the idea of a point-particle breaks down if we approach so closely as to discern its finite size. What such a vortex-particle has in common with a spinning top is this: if a force is applied to it, for example by directing a stream of water against it, then the vortex moves in a direction perpendicular to the direction of the force, and its speed is proportional to the magnitude of the force— again, Aristotelian mechanics with a twist!

You might think that such a strange, counterintuitive dynamics can only arise in complicated systems, such as vortices or tops, because of the large numbers of particles they contain. But no: it can also happen with a single point particle, provided you are in the right limit. The particle must have an electric charge—an electron will do—and there must be a strong magnetic field acting upon it.

According to Newtonian mechanics, an electrically charged particle subjected to a magnetic field is forced to move in a circle in a plane perpendicular to the field. This is a clear-cut consequence of $\vec{F} = m\vec{a}$, combined with the fact that the magnetic force is perpendicular to the velocity and to the magnetic field, and proportional to both. So one possibility for the electron would be to stay where it is—zero velocity implying zero force. But, if it moves, it must move on a circle, and the number of turns per second (also known as 'frequency') will be directly proportional to the strength of the magnetic field. So, if the magnetic field is very large, the electron will go whizzing around the circle. Do you see the beginning of an analogy with the spinning top? No matter how fast the electron is going, the centre of its circular orbit, this abstract mathematical point we call 'the centre of the orbit', remains in a fixed position. In much the same way, the spinning axis of the top would never change its orientation, were it not for the force of gravity that pulls it down.

Now let us see what happens when we apply a force to our electron, for example a pull to the right (Fig. 19). Naive intuition suggests that the centre of the orbit should accelerate in the direction of the pull. But naive intuition fails, because the centre of the orbit is not a material point but an abstract geometrical property of the orbit. What happens instead is that the centre of the orbit begins to move with constant velocity in a direction *perpendicular* to the force—the electron continuing to spin around all along. And one more thing: the velocity of this motion, while directly proportional to the pulling force, is completely independent of the mass of the particle: the centre of the orbit of a proton, about 2000 times more massive than the electron, would drift at

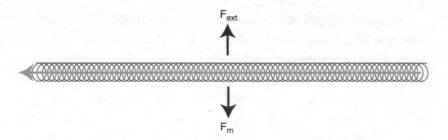

FIG. 19 The steady drift of an electron subjected to a force in the presence of a strong magnetic field perpendicular to the page. The grey line depicts the trajectory of the electron: a tight circle whose centre drifts to the left at constant speed. On a sufficiently long time scale the fast rotational motion averages to zero and drops out of sight; we are left with a motion at constant speed along a straight line perpendicular to the external force F. The *average* force exerted by the magnetic field on the electron (F_m) cancels the external force, so the net force is zero on average.

exactly the same speed, but in the opposite direction. How can all this follow from Newton's equations of motion?

It is, once again, a matter of limits. Recall that the rotational motion of the electron is extremely fast at high magnetic field. By comparison, the centre of the orbit is in slow motion: the electron may go around millions of times before the position of the centre moves by the smallest distance. This suggests that we look at our system in the limit in which the rotational motion is infinitely faster than the drift motion, indeed so fast that it can be ignored: it averages itself out of our observational window. In this limit the position of the electron can be identified with the position of the centre of its orbit, not because the electron is actually there (in fact, it is never there!), but because the average value of its position is there. In the time-averaged world that this limit opens up, we ignore the fast rotational motion as well as the force that is responsible for it ('the dust of inessential facts'), and concentrate on the relevant motion—the slow drift of the centre of the orbit, now identified with the particle. The speed and direction of this drift motion are such that the average force exerted by the magnetic field on the particle exactly cancels the external force. The absence of acceleration in the drift is consistent with the fact that the average net force is zero: Newton would have no reason to complain.

Now that we have understood the motion of an electron in a strong magnetic field we could go back to the spinning top and propose an explanation for the 'laws' of its motion based on the following correspondence: direction of the axis of the top = centre of the orbit of the electron, gravity = external force, precessional motion = drift motion, etc. We have an example of what in literature would be called a metaphor, and there is no reason why it shouldn't be called a metaphor in physics too. We may be talking of one thing or the other, it doesn't matter which: to our intuition they have become one and the same thing.

Let me restate the essential point of this long and complex chapter. I have tried to illustrate the process through which limits open up ideal spaces in which concepts and laws are deployed. Going to the limit means, among other things, discarding a huge amount of realistic detail, so that the essential can be seen. The paradox of the ideal world created by a limiting process is that it is at the same time finite and infinite. Finite is the extent to which it overlaps the real world, infinite are the possibilities it offers to thought. Such a world is like an infinite plane extending in all directions, yet touching the obscure body of reality at just one point. Different limits define different planes, and the totality of these planes constitutes a multifaceted, jagged, discontinuous approximation to the mystery that underlies them all.

Laws are made to be broken

Law and order

The idea of law is inseparable from the idea of order, and order appears, most obviously, as *symmetry*. Look at the portrait of Emperor Tai Zu (Fig. 20), the founder of the Qing Dynasty of China, the Son of Heaven, the absolute ruler whom the lonely librarian of Borges' poem[1] remembers nostalgically as 'the emperor whose serenity was reflected by the world, his mirror, so that the fields yielded their crops and the torrents respected their boundaries'. It is hard to imagine, looking at his portrait, that the man was once a soldier in the battlefield, that his face was soiled with sweat and blood, that he was shouting arrogant orders, riding a black horse, inciting fighters to win or die, that his soul was distorted by greed and hatred, that his every thought and

---

[1] J. L. Borges, *The Guardian of the Books*.

FIG. 20 Portrait of the Qing Emperor
Tai Zu (reign: 1616–1626).

action was aimed at overturning the laws of the land. Yet this is precisely
what a leader had do to rise to power, found a dynasty, and establish a
new order. Order does not arise from symmetry, but from the breaking
of it. Impermanence is its original sin. And because it is born of
impermanence, it is right that it should also die of it.

The notion that order arises from the *breaking of a symmetry* may be
quite surprising at first, but it makes perfect sense when you think about
it. Symmetry means that a figure or a pattern remains unchanged when
subjected to a certain transformation, such as a rotation or a reflection.
For example, the portrait of Tai Zu is unchanged by a reflection about
the central vertical line, and the snowflake of Fig. 21 is unchanged by a
rotation of 60° about an axis perpendicular to the page. The snowflake
looks symmetric to the eye, but it is not nearly as symmetric as the
droplet of liquid water from which it crystallized. A liquid droplet
looks the same in all directions, so it remains unchanged under *all*
rotations, not only 60° rotations. In this sense a water droplet is far

FIG. 21 The structure of a snowflake is invariant under a 60° rotation.

more symmetric than a snowflake. At the same time, there is little doubt that a snowflake is more ordered than a water droplet. In a droplet the water molecules are free to roam over the available space. In a snowflake the position of a single molecule pinpoints the position of all the others, so you cannot move one molecule relative to the others without breaking the structure. Undoubtedly, the snowflake is more ordered than the water droplet, but the price of increased order is reduced symmetry.

You might object that the water droplet, being so disordered at the molecular level, has no symmetry at all: there is no transformation—rotation, reflection, or otherwise—that is guaranteed to preserve a random instantaneous arrangement of the molecules. This is, of course, true—and it might as well be said of any ordered state of matter—our snowflake, for example—not to mention Tai Zu. For even in a snowflake the molecules are not exactly fixed in position: in reality they keep moving and wiggling all the time, and the pattern of fixed relative positions I alluded to in the previous paragraph emerges only if we average over sufficiently long times. So symmetry is absent in the microscopic world at any given instant, but emerges when we look at the system over a sufficiently coarse timescale, disregarding small and rapidly varying fluctuations. Symmetry requires persistence in time.

What makes order so mysterious is that it appears to emerge without an external cue or design, usually from something that was much less ordered, and ultimately, going back in time, from something that was not ordered at all. Where is the organizing principle? Consider, for example, the formation of ice from water. We can assume that the water molecules obey the laws of classical Newtonian mechanics. But we have seen that the principle of inertia, on which Newtonian mechanics is based, is a declaration of symmetry—an affirmation of the uniformity of space. What could be more symmetric than a space in which every point is equivalent to any other? So we might expect that every stable state derived from Newton's laws of motion in the absence of external cues should have the same symmetry as the underlying Newtonian space, i.e. be *homogeneous* (same everywhere) and *isotropic* (same in all directions). In other words, it should be a *liquid*. How do we explain crystals then? How can a crystal form in the absence of a pre-existing template? How does it manage to break a law which mandates uniformity?

Going to the limit helps us to make sense of a seemingly impossible situation. We have already seen that different limits of a physical law, such as Newton's $\vec{F} = m\vec{a}$, can produce vastly different behaviours— Aristotelian dynamics or the dynamics of fast-spinning tops. Now we are going to meet another type of limit, which allows us to understand the existence of objects that break the symmetry of the physical laws. Breaking the symmetry of a law is not the same as violating the law. Rather, it is a transcendental event, which takes us into the very heart of the law, to the point where its ideal validity is most clearly perceived.

Broken symmetry

On the very first page of *One Hundred Years of Solitude*, Gabriel García Márquez introduces two outstanding examples of broken symmetry: ice and the magnet. The magnet, presented to the inhabitants of the village of Macondo as the 'eighth wonder of the learned alchemists of Macedonia', is dragged from house to house causing great sensation: pots and pans tumble down from their places and beams creak from the desperation of nails and screws trying to get free. All this hubbub is, according to our current understanding, caused by the spinning motion of a huge number of elementary particles known as electrons, which swarm inside the magnet.

Electrons are things as abstract as anything you have never seen in your life. At the end of the previous chapter we met them in their traditional role of point-like particles carrying an electric charge. But the whole story is much more complex, for they are also 'spinning tops'. How can a point-like particle—a particle of zero size—be a spinning top when it doesn't have a body that can spin? Evidently the expression 'spinning', as applied to electrons or to any other particle of zero size, must be understood as a metaphor. The idea is that electrons somehow manage to behave as if they were spinning (we'll see how in a later chapter), and because they are electrically charged they create a magnetic field, which in turn produces the strange effects of attraction and repulsion we have all wondered about in our childhood, perhaps thinking they had something to do with love and hate.

In order for the magnet to behave like a magnet, however, one very important thing must happen: namely, a majority of the electrons must agree to spin, most of the time, in the same direction.[2] The ideal situation is illustrated schematically in the lower part (B or C) of Fig. 22. The axes of the spinning electrons (let us call them simply *spins* from now on) are

---

[2] The direction of a spin is simply the direction of the axis around which the rotation takes place. Along this axis the spin points 'up' if the rotation is counter-clockwise, 'down' if it is clockwise.

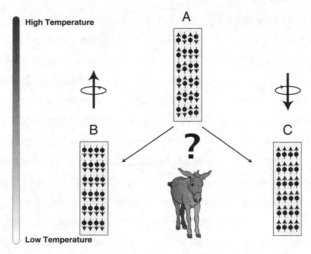

FIG. 22 (A) In a magnet at high temperature the spin orientations are random (for simplicity only two spin orientations are shown). (B)–(C) When the temperature is lowered below a critical value, the electronic spins align in a common direction. How does the magnet choose between the equivalent orders B and C?

oriented along a common direction, parallel to the surface of the bar magnet. By contrast, in part A of the figure the spins are oriented in random directions, of which I show only two, up and down, for simplicity. In reality, the spins are never perfectly aligned. Even in the ordered states, B or C, some spins may be oriented sidewise or even opposite to the majority, and, furthermore, their orientations keep changing with time: but on the whole, at any moment, many more must be pointing along the common direction than against it.

How can such an orderly state be formed *in the absence of external cues*? If you think of space being the same dull background at all points and in all directions (Newton's great abstraction) there seems to be no special reason for the spins to agree to point up, or down, or in any other direction.[3] Like Buridan's ass—the legendary ass who starved to death

---

[3] One might counter that the surface of the magnet creates a preferential direction, with spins parallel to the surface being favoured over spins perpendicular to it. But even this shrewd observation does not help our magnet to choose between the two equivalent orders B and C of Fig. 22.

between two heaps of hay, unable to choose one or the other—our magnet should hesitate forever between different orientations, and thus remain pointless (Fig. 22). But magnets occur as a matter of fact, and facts are obstinate things. How do the electrons decide to spin all together in the same direction? You might say: perhaps they follow a leader, perhaps they hold elections, perhaps they listen to God. Well, that is fantastic enough, but not precise enough. You might as well throw in the towel and say that you don't care to know how they do it. A better approach is to make as few assumptions as possible, i.e. continue to regard the electrons as inanimate particles driven by neutral physical laws. These laws do not lean to the right or to the left, but they do give us a scenario to work with. So here is how the story goes:

> Neighbouring spins have a tendency to spin in the same direction, much as people in the same home or workplace tend to think in similar ways.[4] This tendency, however, is not sufficient to establish a global order. Different groups of people (spins) might choose to abide by different sets of conventions (different spinning directions) and be quite happy in each other's company, even though they'd start barking and snarling as soon as they saw a member of another clan. But, to complicate things, every individual has an irrepressible aspiration to freedom, an innate tendency to think out of line, to explore all possibilities. In the language of spins this is known as 'thermal agitation' and tends to randomize the spins, to knock them out of comfortable alignment.
>
> What controls the intensity of thermal agitation is a number known as *temperature*. At high temperature the thermal agitation is very strong, and the regions of uniform magnetization (groups of spins pointing in the same direction—see Fig. 23) are small and short-lived. As the temperature decreases these conglomerates of spins pointing in the same direction grow larger and live longer, because there is less and less thermal agitation to contend with. While the ordered group holds together, it is tempting for spins outside the group to 'join the club' and start spinning in the same direction. The larger the group, the larger its

---

[4] The reason for this tendency (for spins) is buried deep in the theory of motion known as quantum mechanics. Two electrons spinning in the same direction have a nice way of keeping out of each other's way, thereby greatly reducing the effectiveness of their natural repulsion.

FIG. 23 Magnetic order begins with the formation of cool patches within which all the spins point in the same direction. Size and persistency of the patches increase as the temperature is lowered.

boundary surface, the higher its power of attraction. However, there are still many groups forming simultaneously in various parts of the system, and they may be pointing in different directions, so the global orientation is still undecided. Lower the temperature a little more and something spectacular happens: one of the ordered groups begins to grow aggressively and conquers the whole system. Small pockets of resistance, groups of spins that were initially spinning in different directions, are swallowed up by the voracious conqueror. Isolated spins rush to join the prevailing orientation, much as ordinary people start supporting a war when everybody else around them does. Now what is left is an ocean of spins, all pointing in the same direction. Well, not quite... There are small islands of dissent in this ocean of conformity (Fig. 24). They are quite similar, if you think about it, to the islands of order that existed at high temperature, when chaos reigned. But that was long ago... the insurgents are few and far apart and lack a global plan. Their rare and short-lived demonstrations don't pose a serious threat to the established regime.

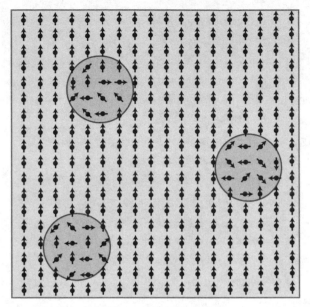

FIG. 24 Even the most ordered state includes small areas of 'dissent'. The disordered patches in this highly ordered magnet mirror the ordered patches in the disordered magnet shown in Fig. 23.

So goes the parable. It is quite credible, but still does not answer the most important questions, which are: how did this particular order come to dominate over other equally possible and, a priori, equally likely ones? What gave it the decisive advantage? How could the fundamental symmetry of space—the equivalence of all directions—be so blatantly broken by a piece of inanimate matter? To approach these questions we must take a bolder step into the abstract.

Permanence is but a word of degrees

A time-honoured adage maintains that 'Everything that can happen will happen'. It is hard to imagine a test for such a vague statement, but it is fair to say that nothing we know contradicts it. A system, initially prepared in a non-symmetrical state (all spins pointing in a certain direction), will in the course of time explore all the possibilities which are open to it, and which do not violate the laws of nature. In the case of a magnet, this means that the ordered state attained at the end of the last section is not the end of the story. True, the spins are permanently locked together. But the global orientation of the spins can still change, and, given sufficient time, it *will* change, so that, in the very long run, the full symmetry of the physical laws will be restored. According to this view, the only truly permanent things in the world must be spherical because only a sphere is equally distributed in all directions. Everything non-spherical is *ipso facto* non-permanent. Furthermore, the sphere must be infinite, because the existence of a finite sphere implies the existence of a special point, the centre of the sphere, and we know there are no special points in Newtonian space. This fearful sphere, *whose centre is everywhere and circumference nowhere,*[5] is the only eternal thing: everything else is temporary.

This sobering conclusion may well be true. As Emerson wrote in the beautiful essay *Circles*: 'permanence is but a word of degrees. Our world seen by God is a transparent law, not a mass of facts. The law dissolves the fact and holds it fluid.' It is amazing how well these words capture the tension that exists between the highly symmetric laws of physics, on one side, and the rigid, crystalline appearance of the world on the other. At the heart of the matter is an issue of timescales or, more precisely, as we will now see, the issue of the relative order in which two limits are taken.

It is very important to appreciate that all the objects that break symmetry (snowflakes, magnets) are *big*, in the sense that they contain an astronomically large number of more elementary particles—for example atoms. If we call N the number of such particles, then the object is well defined in the limit in which N is very large or, as we like to say, N tends to infinity. On the other hand, if we want to assess the stability of a given order we must watch it for a long period of time T:

[5] B. Pascal, *Pensées*, 199 (72).

ideally T must tend to infinity too. Now the issue is the relative order in which these two limits, N tending to infinity and T tending to infinity, are to be taken. On an infinite timescale—the scale of eternity, available only to God—T tends to infinity before N. Then the laws of physics hold sway and everything must be perfectly symmetric. Opposite to this, on the ridiculously short but tremendously important timescale of our historical record (T finite), large objects appear to be frozen in permanent low-symmetry structures, and the laws of physics are, in a sense, suspended. So we can say that broken symmetry emerges when we let the number of particles in a system, N, tend to infinity, while keeping the timescale T fixed at some large value (T is allowed to tend to infinity at a later stage, after the limit of N tending to infinity has established the broken-symmetry state). But, if T is allowed to go to infinity first, while N is still finite—this is presumably the point of view of God—then the broken symmetry is dissolved by the Law. 'Whatever dies was not mixed equally'—writes the poet John Donne in *The Good Morrow*, undoubtedly alluding to the transient nature of broken-symmetry states.

Let us try to understand how these ideas work in the case of a magnet. The first thing to absorb is that it doesn't take any long-range planning on the spins' side to settle into an ordered state. Each spin is only aware of its immediate neighbours: it has no way of knowing what is going on far away. What happens at the critical temperature is that our system of short-sighted, randomly oriented spins, develops a strange property, which can be described as an *infinite sensitivity to the smallest external influence*. The slightest hint to go this way or that way—a hint so slight that it would have no appreciable effect in ordinary circumstances—is amplified with dramatic consequences. This slightest hint might be a stray magnetic field (for example the magnetic field of the Earth), which gives a slight advantage to groups of spins that happen to point in the direction parallel to the magnetic field. Immediately, those groups of spins begin to grow disproportionately, at the expense of other groups, and within a very short time all the spins are pointing up along the magnetic field.[6] The key point is that the predisposition to order—the infinite sensitivity to external hints—emerges only in the limit of a very large system, i.e. in the limit of N going to infinity. At finite N the system will have a finite (perhaps very large, but finite) sensitivity, which means that a magnetic field will not be able to pinpoint the magnet unless its strength exceeds a certain threshold. It is only in the limit of infinite N that a global alignment can be induced by an arbitrarily small external field, and it is only when this happens that one can say that the symmetry is spontaneously broken.

Once the order has been established, you can turn off the infinitesimal cause that precipitated it. It is here that the magic really shows up. One might think that removing the external field would cause the magnet to revert to its original, undecided state. But that doesn't happen—not as long as the temperature stays below the critical value. The spins are now locked in a common orientation and cannot get out of it without a concerted effort, which would require long-range planning—not available. To be sure, if one waits long enough, then sooner or later it will happen that all the spins turn together in a different direction, because 'everything that can happen will happen'. But the waiting time for this momentous event is stupendously large. With N running in the trillions of trillions even for the tiniest speck of matter, the waiting time dwarfs the age of the Universe: so the re-orientation will never happen, as far as you and I and all the foreseeable generations of our descendants are concerned, and the law can be said to be broken, if not in form in substance.

Why does it take so long to go from one orientation to the other? This is, I think, a very important question. Because order is usually beneficial to life, we tend to focus on the difficulty of attaining it, but

[6] You might object that, admitting such an external field is equivalent to breaking the symmetry of space. This is true, but you must imagine that the external field is turned off immediately after accomplishing its goal—orienting the magnet. Then you are left with a genuinely broken symmetry state in a perfectly symmetric space.

the fact is that it may be even harder to get out of it. The hardest thing of all is to replace one kind of order with another. Suppose all the spins are pointing to the North, and you, a single individual, have this sudden inspiration that it would be better to point to the East. You can of course show everyone the way by turning yourself in that direction, but the only thing you are likely to accomplish is to make yourself uncomfortable, and your neighbours angry. In the most optimistic scenario, you'll be considered a fool or a madman—more likely, a dangerous criminal to be eliminated as quickly as possible. A few kindred souls who saw you from a distance may choose to follow your example, and will likely suffer a similar fate. And yet it would be so easy to turn all together, if all chose to do so simultaneously as part of an organized movement. No one would suffer, and the final product of the revolution might even turn out to be disappointingly similar to the starting point. But such a concurrence of individual spin-wills is so fantastically improbable that it will never happen in the lifetime of the universe. Change of order is inhibited by the fact that the operation is extremely risky for the few initiators, even though it might be ultimately beneficial to many followers. The fact that revolutions occur far more often in human societies than in inanimate systems demonstrates the decisive role played by consciousness, with all the abstract notions of beauty, truth, and justice that come with it.

Broken symmetry is so pervasive that we take it for granted. The regular structure of an ice crystal does not surprise anybody. It seems quite natural that the atoms should arrange themselves in a pattern of relatively fixed positions, even though the symmetry of this pattern is infinitely lower than that of the underlying space. Similarly, the observation that a water droplet has a surface does not elicit cries of surprise, even though the very existence of that surface is a consequence of broken symmetry. In some cases the broken symmetry is quite abstract. A particularly subtle example is provided by the phenomena of *superconductivity* and *superfluidity* in matter. Superconductors are ordinary metals which, below a certain critical temperature, lose their electrical resistance altogether: that is to say, they can keep an electric current flowing forever without the help of a battery. Similarly, liquid helium at very low temperature becomes truly unstoppable: it creeps through the tiniest aperture and, under appropriate conditions, can climb out of a container by its own means. These dramatic behavioural changes are not accompanied by any visible reduction in symmetry. And yet some kind of order sets in at the transition and some kind of symmetry is broken. What is it? Physicists call it *gauge symmetry* and the resulting order *off-diagonal long-range order*. The difficulty with these concepts is that they make sense only within the context of the quantum theory of matter, and cannot be explained by analogy with the behaviour of objects in the visible world. This is not to say that they cannot be explained in simple and intuitive terms, but the necessary intuition is acquired only with considerable effort. An attempt to explain super-conductivity will be made in Chapter 11.

The uniformity of time—that fundamental symmetry which, as we are going to see in Chapter 8, is responsible for the law of conservation of energy—is another symmetry that is spectacularly broken in every-day life. Indeed, life itself is a process through which a steady input of energy from light and food is converted, through myriad low-key processes, into a complex drama of birth and death, which breaks the monotony of time. The very presence of 'I' at this particular instant, out of millions of years during which I could have existed, is a sort of miracle of broken symmetry. Blaise Pascal must have had that in mind when he wrote:

> I see the terrifying spaces of the universe hemming me in, and I find myself attached to one corner of this vast expanse without knowing why I have been put in this place rather than that, or why the brief span of life allotted to me should be assigned to one moment rather than another of all eternity which went on before me and that which will come after me.[7]

[7] B. Pascal, *Pensées*, 427 (194).

Like every broken symmetry, individual consciousness comes with an illusion of eternity. But it is not spherical; and the basic laws demand its demise.

Circles

We now begin to see in what sense laws can be broken while still being obeyed. The idea is this. Suppose you set for yourself the immodest goal of explaining all of reality in terms of a certain set of fundamental objects, such as particles, vortices, fields, strings, or whatever your imagination can supply you with. You throw your basic objects in space and time as actors on the stage, and declare that they obey a certain set of laws—the script. If the laws are well designed and rigorously obeyed, you should see that they lead to the formation of a larger circle of composite objects, such as atoms and molecules. These composite objects break the original law in the following sense: they have, in general, lower symmetry than the laws under which they are constituted. They also have new properties that did not exist in the original script. For examples, atoms and molecules have a mass—a key concept in Newtonian mechanics—while the fundamental fields from which they emerge are believed to be massless or nearly so. Although atoms and molecules, unlike fields and truly elementary particles, are non-permanent (they eventually disintegrate as time goes by) they are nevertheless long-lived, sometimes extremely long-lived. Thus, provided you are careful to take the right limits in the right order, i.e. to look at your 'transient objects' with the appropriate spatial and temporal resolution, you are perfectly justified in using them as elementary building blocks of a higher-level world—the world of atoms and molecules.

You now need to discover the laws that govern the behaviour of atoms and molecules on their own scale. These laws don't have to be similar to those of the lower level, even though they cannot run *against* them. And this is why condensed matter physicists and chemists, whose job is to formulate the laws of the atomic and molecular world, usually have little to say to elementary particle physicists, who build those atoms and molecules from more primitive blocks.

The process I have just described can be repeated indefinitely. The elementary objects of each level are constituted under the laws of the previous level, but obey the laws of their own level. From the molecules of chemistry one assembles the living cells of biology, and from the living cells of biology one forms plants, animals, and human beings ready to be studied by an army of physicians and psychologists; and from humans beings one forms societies, of which sociologists and economists study the laws; and so on, circle after circle, to higher and higher levels of complexity. We have then, quite evidently, a recursive structure of laws. Nor is it easy to see an end to the spiral of successive generalizations. For even if Nature got tired of supplying new materials, there would be our own thoughts to provide food for new thought. This is precisely what happens in the arts, and especially in literature, where new worlds are created by fantastic precision, and literature itself becomes the subject of more literature.

The idea of a hierarchical organization of laws in the realm of natural science was put forward by a condensed matter theorist, Philip Anderson, in a paper he published in 1972 under the catchy title *More is Different*.[8] In those days elementary particle physics was widely regarded as the cutting edge of science, and its practitioners felt they were the only scientists doing truly fundamental research (they still do). Stung in his pride, Anderson—who later won the Nobel prize for physics—set out to confute this view. He argued that not only condensed matter physics, but every branch of science has its own set of fundamental laws. While these laws cannot run contrary to the laws of the underlying levels (e.g. the laws of chemistry cannot contradict the laws of physics, and those of biology cannot contradict the laws of chemistry and

---

[8] P. W. Anderson, *Science* 177, 393 (1972).

physics), it is nevertheless impossible, in practice, to derive the former from the latter. And the discovery of higher-level laws requires as much creativity, originality, and genius, as the discovery of the more basic laws. Anderson brought up, as a case in point, the principle of broken symmetry—an example that he, as one of the leading theorists of superconductivity and magnetism, was thoroughly familiar with. But broken symmetry, as we have just seen, emerges as a sharp feature of reality only when you put together more and more particles, i.e. when you go to the limit in which the number of particles tends to infinity. This is why 'more is different'.

The title alludes to an exchange between two American writers, Ernest Hemingway and Francis Scott Fitzgerald. According to legend, Fitzgerald—who had a childish admiration for the lifestyles of the rich and famous—asked Hemingway: 'Don't you think the rich are different from the rest of us?'—to which the more manly author replied: 'Yes, they have more money'.

Obviously Hemingway was right, and yet his quip does not take anything away from the truth of Fitzgerald's observation, which cuts to the heart of the matter. Think, for example, of the life of Drew Preston as described in E. L. Doctorow's novel *Billy Bathgate*. At age 20, Drew was already married and separated from her husband, was mistress to a famous gangster who had murdered her previous boyfriend, and was in danger of being murdered herself. And yet...

> she knew about sailing and oceans too, and beaches to swim from with no crowds on them and...in fact all the pleasures of the planet, all the free rides of the planet...This was what wealth was, the practiced knowledge of these things so you could appropriate them for yourself.

New worlds demand new laws, which define new objects and create new possibilities. Nowhere is this more evident than in the development of great fantastic novels. So, in one episode of *One Hundred Years of Solitude*, flowers fall from the sky to honour the death of a great man; in another, the police fail to see the man they are chasing—for he has become invisible to them. These events, which dreary common sense declares impossible, become possible within the tangential, yet exact reality of the novel. And a virgin can give birth to a baby, and a man can rise from the dead.

5

Of first and last things

An issue of control

Apropos coming back from the dead: what about that annoying feature of reality that makes this feat so utterly impossible? I am talking about *irreversibility*, of course.

Imagine yourself riding a motorbike to a fun destination—a party or a date. You are confident and in high spirits, your mind roams over the past and the future, you feel in control of your reality. All of a sudden you hear a crash, your muscles tense up, your head starts spinning—a moment later you are lying flat on the road, next to the wreck that was your bike. People approach you from all sides: are you OK? Hurt and confused, you try to minimize the incident, almost to deny it: you are fine, it's only a scratch, you are happy to be alive. But your destroyed bike, the police, the ambulance, are evidence that you have not been dreaming, that something has happened which cannot be undone.

If you have ever been in a road accident you probably understand the sense of unreality that seizes the mind in the face of irreversible events. Normally, we don't take steps that cannot be retraced, and we become nervous when a door locks behind us. In the world of thought, reversibility is the rule. We can undo fantastic connections or logical inferences without penalty. Years ago I wrote a cerebral story in which a man vowed to take revenge on his rival by dreaming up his death so accurately as to make it real. The extravagant task takes the best part of his life (which could have been put to a better use), and when he finally succeeds he knows that he has succeeded because the dream, which had been under control until then, suddenly acquires an independent life. The new-born reality hastens to its conclusion, leaving the mind of its creator at once empty and sad.

There are two essential differences between reversible and irreversible actions. The first is that the former do not exist in nature but only in our mind: they are idealizations of real processes. The second is that these ideal processes are tightly controlled while real ones are not. Reversible processes do not leave a trace in the world other than what is strictly intended. They usually involve something very simple such as pushing a switch from the 'on' to the 'off' position during a power outage. Because the action is inconsequential (there is no power), we can undo it without anybody noticing that we ever did it. Irreversible events, on the contrary, are poorly controlled and have a lot of unintended consequences, which get permanently imprinted in the surroundings. Think of the explosion of a bomb, triggered perhaps by the same switch, now connected to an electric circuit.

Because of their messy consequences, irreversible actions cannot be undone—not completely at least. So, for example, I might be sitting in my home pondering a momentous decision: shall I or shall I not declare my love for Yvette? I consider the question from all points of view. I rehearse my lines many times and try to imagine her response. This is an exercise in fantastic precision. In each experiment I stretch reality one way or the other, examine the outcome, and return to the starting point: the deformation is elastic, like stretching the string of a bow. Characteristic of this mode of thinking is that I remain free, uncommitted. Up to the very last moment, when I climb the stairs that lead to her apartment, knock on the door, perhaps hoping that it will not open, I have the freedom to retrace my steps. Should I choose to do so, neither she nor anybody else would ever know how close I came to declaring. But the door opens and I go in, and I speak. The arrow is released, it flies through the air with a frightful whiz. There is no going back. The fact that I love her and want to marry her is firmly planted in her mind. It will propagate, it will be retold, it will affect the lives of others, and perhaps it will crystallize in a church or in a court of law, in front of authorities, witnesses, and guests. A chain of events has been started, which amplifies to social consequences what started as a thought experiment in my head. I have little control on this chain of events. The more information is released, the less control I have. This is the essence of irreversibility.

The Second Law

What I wrote in the last few paragraphs is not quite the way the story is told in physics textbooks. In order to achieve mathematical precision, physicists consider only very simple situations, which can be analysed more or less completely: for example, a gas in a box. Let us say that we have a box partitioned into two chambers, as shown in Fig. 25: the left chamber contains the gas, the right one is empty. I must stress that in drawing this figure I have already adopted the modern picture of the gas as a collection of particles—molecules or atoms—which obey the laws of mechanics. This mechanistic picture was far from evident to physicists and chemists just about a hundred years ago. Indeed, as we will see later, the very fact of irreversibility was, to many scientists, irrefutable proof that any mechanical description of matter must be in error. But let us proceed in order.

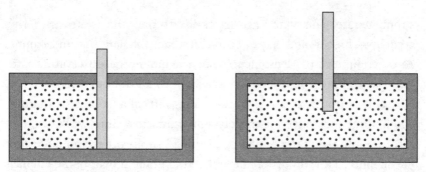

FIG. 25 A textbook example of irreversibility. Shortly after the partition is lifted, the gas fills the whole box and will never spontaneously reassemble in its left half.

Let us lift the partition that separates the two halves of the box: it doesn't take much imagination to predict that the molecules will spread out until the box is uniformly filled with gas. Indeed this happens very quickly, and not because the molecules dislike each others' closeness—as people probably would—but because there are many more ways to fill the whole box at random, than to remain confined to half of it. Things that can happen in many random ways have a tremendous advantage over things that can happen in only one or a few ways. You can lose a lottery with almost every number but you win it only with one number. Error always has a huge advantage over truth—so has sin over virtue. Similarly, the probability of the gas filling the box is overwhelmingly larger than the probability of it remaining confined to the left half of it. Furthermore, once the expansion has taken place there is no going back: the gas will never—and we will come back later to the meaning of that 'never'—retreat of its own accord to the left chamber. In brief, it is impossible for the gas to *spontaneously* reassemble in the left half of the box.

There is something puzzling about this impossibility. Notice that the expansion of the gas takes place under the laws of mechanics, which are perfectly reversible in time. In other words, if a certain motion is allowed, then the inverse motion is allowed too. Why, then, should it be possible for a gas to spontaneously expand, but impossible to spontaneously contract? This impossibility must be rooted in a law somehow higher and more powerful than the laws of mechanics. Another hint of the existence of such a law is the observation that heat never *spontaneously* flows from a cold to a hot body. Another one is that it is not possible to extract energy from a source in thermal equilibrium and convert it entirely and *systematically* into work.[1]

These impossibilities gradually dawned on physicists and engineers of the nineteenth century as they reflected on the power of the engines that were propelling the industrial revolution. In those days, heat had already been successfully interpreted as a form of energy—a concept well rooted in Newtonian mechanics. There was a consensus that energy is neither created nor destroyed, but can be freely exchanged like money on the market: this is the content of the First Law of Thermodynamics. And yet everyone could see that the mere conservation of energy would not have prevented heat from flowing spontaneously from a cold body to a hot body. There had to be more than just the conservation of energy. There had to be a Second Law of Thermodynamics to account for the impossibility of so many plausible transformations.

The decisive step was the realization that the Second Law could be expressed as a *law of increase of entropy*. What does this mean? The word 'entropy' comes from the combination of the Greek *energeia* (energy) and *tropy* (transformation) which suggests the possibility of a change in the *quality* of energy. In other words, energy is not just like money—a universally exchangeable quantity—but more like a commodity (say, bread) which derives its value from the specific form in which it is presented. The same quantity of energy is far more useful when it is stored in an orderly structure (e.g. in the rotation of a windmill's wheel,

---

[1] The adverbs *spontaneously* and *systematically* are essential to the correctness of these statements. It is quite possible to force heat to flow from a cold to a hot body (a refrigerator does precisely that): the point is that the flow will not happen spontaneously. It is also possible to extract energy from an equilibrium source, just as it is possible, with some luck, to win money at a casino. The problem is that it cannot be done systematically.

or in the chemical bond between carbon atoms) than when it is distributed over many chaotic and uncontrollable motions (e.g. in the exhaust gas released by a combustion engine). Entropy is an inverse measure of the quality of energy, of its usefulness: a given quantity of energy at high entropy is less useful than the same quantity of energy at lower entropy. A given amount of energy in a flame is more useful than the same amount of energy stored in a pot of water at room temperature, and it carries correspondingly less entropy. When heat flows spontaneously from a hot body to a cold body the energy goes from a more useful form (low entropy) to a less useful form (high entropy). Thus, the entropy increases at the same time that the quality of the energy is degraded. The inverse process—heat flowing spontaneously from the cold to the hot body—would entail a decrease in entropy, and therefore is never seen in nature.

For an ideal gas in equilibrium one can show that the entropy is completely determined by the temperature and by the volume of the container, and increases when either of these two quantities increases. So the free expansion of Fig. 25, with the temperature remaining constant and the volume doubling, is accompanied by an increase in entropy. The inverse process, which would reduce the entropy, does not occur in nature. This and many similar observations find an elegant explanation in the Second Law, according to which

> The total entropy of a closed system (a system that does not exchange heat with the external world) can never decrease.

An immediate consequence of the Second Law is that the total entropy of the Universe, regarded as a huge closed system, can never decrease. In fact, it will always increase, except in reversible processes, which do not change the total entropy. For, if a reversible process increased the entropy, then the inverse process would decrease it, violating the Second Law. Another consequence is that there is a limit to the efficiency of a thermal engine: you cannot extract heat from a source in equilibrium (e.g. a boiling pot of water) and transform it

FIG. 26 In this apparent reversal of the free expansion of Fig 25, a partition is slowly moved from the right to the centre of the box, trapping the gas on the left. However, in order to truly restore the initial state of Fig 25, some heat must escape from the box, causing the entropy of the environment to rise more than the entropy of the gas decreases.

entirely into useful work for, if you could do that, you would decrease the entropy of the world. One more consequence: no engine can outperform the ideal reversible engine, i.e. the engine that generates exactly zero entropy. It follows that all ideal reversible engines have exactly the same efficiency. Otherwise, you could find one that outperforms another—thus achieving what we have just declared to be impossible.

On the positive side, the Second Law does not prevent us from reducing the entropy of a system that is not 'closed', for example a system from which we can extract heat—which is what we do when we cool our home with an air conditioner. But, even in this case, the Second Law has something important to say. True, by extracting heat you can reduce the entropy of your home by some amount, but the extracted heat will have to be released into the environment, where it will cause the entropy to rise by an even larger amount. This is because the global system—home plus environment—is closed, and its entropy cannot decrease.

It takes some effort to digest this deceptively simple point. Let us go back to the example of the free expansion of a gas. Surely something can be done to force the gas back into the left half of the box, even though this diminishes its entropy. For example, we may slide a partition along the right wall of the box, and move it slowly to the centre of the box. The gas is pushed back to the left compartment, as shown in Fig. 26. This is easy enough, but the problem is that, at the end of the process, the gas is not exactly as we want it to be. In fact, it has become slightly hotter, due to collisions between the molecules and the moving partition. It turns out that the increased temperature exactly makes up for the decreased volume, so we haven't changed the entropy after all! To correct for this, we must allow the extra energy to escape the container and be dispersed in the environment. By allowing this, how-ever, we start a chain of uncontrollable events—a chain that extends well beyond the little box of our experiment and, in principle, out to the boundaries of the universe (if there are any). The gas is now exactly as it used to be: but the rest of the world will never be the same—its entropy has increased.

The reality of atoms

What is the relation between the Second Law of Thermodynamics and the laws of mechanics? The Second Law is like a higher authority, which forbids processes that would be perfectly legitimate as far as mechanics is concerned. But how does it do that? Remember, a higher level law cannot run counter to the laws of the underlying level. The difficulty is that mechanical processes can develop equally well forward and back-ward in time. A possible mechanical motion played in reverse remains a possible mechanical motion. Thus, in a purely mechanical world there would be no place for a 'before' and an 'after' like we have in the real world. Irreversibility seems to positively rule out a description of the world based solely on the motion of atoms and molecules.

This general philosophical difficulty is compounded with technical difficulties. The entropy of the Second Law, while mathematically well defined, is not a thing to which one can easily attach a mechanical interpretation. Considerations of probability arise naturally when we try to understand the irreversibility of the free expansion of a gas from a molecular point of view. Entropy is somehow related to the likelihood of a particular macroscopic state. The more likely states have higher entropy than the less likely ones. Then the Second Law tells us that natural processes proceed from less likely states to more likely ones, with the most likely of all states—the equilibrium state—being the final destination of every process. All this sounds vaguely plausible, but how do we connect these ideas on the one hand to mechanics and on the other hand to the known facts of the degradation of energy, the limited efficiency of engines, etc.?

The man who deserves the credit for making these connections is the great theoretical physicist Ludwig Boltzmann. Boltzmann was a man of acute sensitivity, who could be moved to tears by a shade of colour of the ocean.[2] By all accounts he was a generous, fun-loving fellow, who enjoyed machines and manual work. But, in the later part of his life he began to suffer from bipolar disorder, which ultimately drove him to suicide. A poem he wrote during the last period of his life throws some light on his personality. The poem, entitled *Beethoven in Heaven*,[3] begins with Boltzmann's soul escaping the body and ascending to Heaven. There he hears a 'mighty choral hymn', which—the angels tell him—has been composed by Beethoven himself, now master composer in Heaven, 'upon the Lord's command'. Sadly, the hymn does not measure up to Beethoven's earlier compositions. Why? The most powerful source of inspiration is missing: there is no pain in Heaven.

The task Boltzmann set for himself was nothing less than explaining the behaviour of gases in terms of atomic motions—motions that are governed by the laws of Newtonian mechanics. He did this at a time when the reality of atoms was still being questioned from all sides. He realized that the mysterious entropy of the Second Law is essentially the number (actually the logarithm of the number) of atomic states of motion that 'look the same', i.e. produce the same global properties,

[2] See Boltzmann's memoir entitled 'A German Professor's Journey into Eldorado', reproduced in the book by Carlo Cercignani, *Ludwig Boltzmann, The Man who Trusted Atoms*.

[3] F. Rohrlich, 'A poem by Ludwig Boltzmann', *American Journal of Physics* 60, 972 (1992).

at the macroscopic level. This 'Boltzmann principle' is now engraved on his tombstone—although in a form that Boltzmann himself never wrote. From this principle he went on to 'derive' the Second Law—the Law of Increase of Entropy. That is to say, he proved that the number of microscopic states accessible to the atoms of a gas must continually increase as the atoms evolve under the laws of Newtonian mechanics.

I put the word 'derive' in inverted commas because such a derivation is, strictly speaking, impossible. Today, we see the Second Law as an example of an *emergent law*. It breaks the laws of mechanics, on which it is based, without ever violating them. It creates a sharp distinction between past and future (with the future being the direction in which the entropy increases) when no such distinction exists in mechanics. This is a perfect example of the phenomenon described in Chapter 4: a higher-level law breaking the symmetry of the underlying law. What is broken is the equivalence of past and future, and the broken equivalence becomes more and more clearly recognizable as we consider systems containing larger and larger numbers of particles. The Second Law of Thermodynamics emerges from mechanics in a special limit, which was first identified by Boltzmann. The realization of this limit requires, with other more technical conditions, that the number of particles tends to infinity. Once again, an abstract limit turns out to be the key to our understanding of reality.

Maxwell's demon

Because of its philosophical implications, the Second Law is one of the most heavily criticized laws of physics. Initially, the criticism focused on the apparent conflict with mechanics. The problem is that the Second Law makes an absolute statement—the entropy will *never* decrease—while mechanics yields at most a probabilistic statement, an assertion that the entropy increases almost always, but not *always*. The probabilistic statement becomes an absolute certainty only in the limit in which

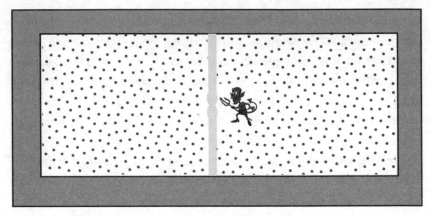

FIG 27 Maxwell's demon's job: to guard the doorway between two gases, so that no molecule exits the left compartment.

the number of particles tends to infinity (and even then, only after excluding some special initial conditions). But since real systems have a finite number of particles, a small uncertainty remains. This residual uncertainty is like an open door through which those who don't fully appreciate the power of limits can carry attacks against the logical structure of the theory.

Another brand of scepticism has led many to question whether the Law is really as absolute as it claims to be, or whether it might be circumvented, perhaps using consciousness as the ultimate weapon. None of the attempts made in this direction is more famous than 'Maxwell's demon', the fantastic creature born out of the restless imagination of James Clerk Maxwell, one of the greatest physicists of all times (more about him in Chapter 7). His objection to the Second Law has a crisp post-modern flavour, so I will discuss it first.

We have seen that a gas in a box will never spontaneously retreat to one half of the box, unless forced to do so by actions that ultimately increase the total entropy of the universe. But what if the agent of the transformation were a creature of microscopic size, one that could operate quickly and accurately at the molecular level without leaving a mark in the external world?

Imagine a little demon standing in front of a microscopic aperture in the partition between the right and left compartments of the box (Fig. 27). Whenever a molecule attempts to escape from the left compartment the demon pushes it mercilessly back; but when he sees a molecule approaching from the right, he admits it promptly. So, after a while, all the molecules are in the left compartment, and the Second Law has been violated, provided, of course, the demon himself has not increased its own entropy or the entropy of the rest of the world.[4]

Now, why should that be the case? All the demon has to do is keep his eyes open, watch for incoming molecules from the left and push them back. It is really a kind of computation in which one says: if this or that happens then do this or that. The computation can be done reversibly, that is to say at zero entropy cost.[5]

The catch—or, at least, one of the catches—turns out to be surprisingly subtle. In order to acquire fresh information about the incoming molecules—and there are billions and billions of them approaching the aperture at any moment—the demon must periodically discard previously acquired and now useless information about events that have already been processed. It is like that kids' game in which a recorded voice issues a random sequence of orders, which the kid is supposed to obey by pushing, pulling, or squeezing a small handlebar as the voice

[4] At first sight, the same task could be accomplished more efficiently by a microscopic trapdoor—a door that opens only in one direction. The problem is that the door, being itself microscopic, will be affected by collisions with the molecules, which will cause it to make 'mistakes', sometimes admitting a molecule from the wrong side, sometimes pushing back a molecule that should have been admitted. In the final state of equilibrium the rate at which these mistakes occur is such that the number of molecules crossing from the right to the left equals the number of molecules crossing from the left to the right.

[5] An ingenious scheme to do this computation without spending energy is described in the article by Charles H. Bennett, 'Demons, Engines and the Second Law', in *Scientific American*, November 1987, pp. 108–116.

commands. The game goes on until one makes an error—and it is clear that this will happen very soon unless the player is able to concentrate entirely on the last command, forgetful of what he heard or did earlier. In other words, one must keep one's memory clear, and so must Maxwell's demon if he wishes to keep his job. It is the process of discarding information that causes the entropy to rise and protects the Second Law. Information is stored in a physical arrangement of neurons and/or transistors. When you (or the demon) do a 'memory clear' this arrangement is destroyed, causing a leak of entropy into the environment. In effect you are using the environment as a waste basket for your no longer useful data. We thus return to our first take on irreversibility: that it always arises from a loss of information, from imperfect control causing our actions to have myriad unintended effects in the big wide world.

It seems to me that the Second Law is, more than any other law of nature, a law of *nature as understood by humans*. In Chapter 2, I argued that every attempt to make sense of the world begins with setting aside some facts that we deem to 'not affect the question'. We must already know what we are looking for, in order to see it. In the case of the Second Law, what would be the meaning of an expression like 'the wine glass is broken and its pieces will never get back together' if we didn't have a clear conception of the unbroken glass as an object from which one can drink? The atoms that constitute a wine glass do not care whether the glass is whole or broken. To them, one arrangement is as special as any other, and there are no wine glasses, whole or broken, to talk about. The hungry wolves of Jack London's novel *White Fang* did not see a human hand as a hand, capable of infinite and subtle skills, but only as flesh to be eaten.[6] Evidently, we must discard huge amounts of detailed information in order to arrive at such elegant, high-level expressions as 'wine glass' and 'broken wine glass'. Both expressions are compatible with a tremendously large number of atomic arrangements, but of these two

---

[6] Jack London, *White Fang*, Chapter 3.

enormities the second—the number of arrangements associated with the concept of broken glass—is overwhelmingly larger than the first, because it does not have to satisfy the constraints of global shape without which it's impossible for a glass to function as a glass. And this is why it is so easy to go from the whole to the broken, but nearly impossible the other way around.

The exception and the rule

Let us go back to the troubled relationship between the Second Law and mechanics. The Law asserts that the total entropy must increase or remain constant. But this cannot always be true, as can be seen from the following argument.

Once again, we start with our gas in the left compartment of the box shown in Fig. 25, lift the partition, and watch the gas expand. This time, however, half way into the expansion we do something tricky, namely, we reverse the direction of the velocity of each molecule. A molecule that was moving to the right will turn back and start moving to the left with the same speed. This may be hard to do in practice, but there is nothing that forbids it in principle, and surely we can design a machine that does precisely this. We call this machine a *Loschmidt's demon*, in honour of Professor Josef Loschmidt—the friend and critic of Boltzmann who first came up with the idea. The Loschmidt's demon does not attempt to violate the Second Law: its purpose is simply to create a special state in which the velocities of the molecules are precisely opposite to what they would have been in the freely expanding gas. Having accomplished this, the demon withdraws and the gas continues to evolve under the laws of Newtonian mechanics. What happens is that the gas retraces its steps, as if the expansion had been filmed and were now being played backwards. It will surely reassemble in the left half of the box, whereupon it will proceed with overwhelming probability to expand again, unless we promptly lower the partition.

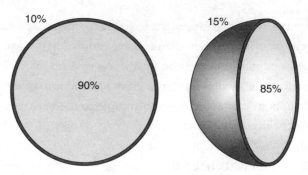

FIG. 28 The distribution of points inside a ball depends strongly on dimensionality. In two dimensions, a 5-cm-thick annulus near the boundary of a disk of radius 1 m, encloses about 10% of the area of the disk. In three dimensions, a 5-cm-thick shell near the surface of a ball of radius 1 m encloses a considerably larger fraction—about 15%—of the volume of the ball. A 5-cm-thick shell near the surface of an N-dimensional ball of radius 1 m would enclose practically the entire volume of the ball, in the limit in which the number of dimensions tends to infinity!

So you see that from a strictly mechanical point of view it is not impossible for the gas to reassemble and for the Second Law to be violated. It is just a matter of starting from the right initial conditions. But it is also a matter of timescales since, in the example considered, the contraction of the gas will soon be followed by a new expansion. One might say that mechanics allows *brief periods of time* during which the Second Law can be violated, due to some very special and soon-to-be-lost correlation between the positions and the velocities of the molecules. These exceptional conditions are so rare, and produce their effects for such short periods of time, that they are simply declared *non-existent* in the idealized world in which the Second Law rules. It is not that the Second Law runs counter to the law of mechanical reversibility; rather, it recognizes that mechanical reversibility is ignorable in the limit of interest.

A simple analogy helps us to understand the sense in which rare events may be considered irrelevant. Consider a disk delimited by a circle in a plane. Looking at the left side of Fig. 28 we get the impression that only a small part of the disk lies near the circumference. To be more precise, let us say that the radius of the disk is 1 m and that a point of the disk is 'near' the circumference if its distance from it is less than 5 cm. Then the probability that a randomly chosen point will be near the circumference (the shaded region in Fig. 28) is about 10 per cent. The same calculation, done for a three-dimensional ball of radius 1 m (Fig. 28, right side), tells us that about 15 per cent of its points are within 5 cm of the spherical surface. This is already distinctly higher than the value we found in two dimensions. Now let us imagine a ball in more than three dimensions. I cannot draw this for you, but I can easily tell you what it is because a word is worth a thousand pictures! An N-dimensional ball is the set of points in N-dimensional space whose distance from a special point called 'the centre' is less than or equal to the radius (1 m in our example). And an N-dimensional space is like ordinary space, except that its points are defined by N coordinates, instead of the usual three. It can be shown rather easily that the probability of finding a point within 5 cm of the surface of the N-dimensional ball increases steadily with N, and tends to 100 per cent— an absolute certainty—in the limit in which N tends to infinity.

This striking result is, in a sense, the foundation of that branch of theoretical physics known as *statistical mechanics*. It tells us that, in the limit of large N, an N-dimensional ball is practically all surface and no volume: it has no depth whatsoever. What this means is that a randomly chosen point within the ball will, with overwhelming probability, fall within a very small distance of the surface. Special points, such as the centre of the ball, may be of great interest to the artist, but from the statistical point of view they are so special as to be insignificant. Picking one of those points by sheer luck would be as unlikely as owning the winning ticket of a lottery in which infinitely many tickets had been issued.

The relationship between mechanics and the Second Law is a little like the relationship between an infinite-dimensional ball and its surface. The infinite-dimensional ball represents the possible states of motion of the atoms. Mechanics must cover them all, including the very special and unlikely. Statistical mechanics, on the other hand, is concerned only with typical behaviours, and from its point of view only the surface of the ball is relevant. This opens up the possibility that some statistical laws might hold for points on the surface of the ball, and yet be false in general.

I should emphasize that what is irrelevant in one context may well become the most relevant thing in a different context. For example, the centre of an N-dimensional sphere may be statistically irrelevant, but, without it, one would not be able to define the sphere. Or, if you are a writer, you might be infinitely more interested in the exceptional behaviour of a single individual—Moosbrugger—than in the run-of-the-mill activities of millions of ordinary people. Likewise, the Second Law must be understood within its own context—a statistical one. It is not 'rigorous', yet the cases in which it fails become increasingly rare as the number of particles tends to infinity.

The same things return

Everything we have seen so far in this chapter suggests that macroscopic irreversibility is a special type of broken symmetry which breaks the equivalence between past and future. It selects an 'arrow of time', just as magnetism selects a preferred orientation of the spins. The analogy can be pushed further. In Chapter 4 we saw that the broken symmetries of space eventually come back with a vengeance, dissolving any ordered state, provided we are willing to wait for a sufficiently long time—possibly longer than the age of the universe. Is there a similar effect that dissolves the arrow of time? Yes, there is, and it is known as *the recurrence paradox*.

The recurrence paradox is based on a deceptively simple property of classical mechanical systems. Consider a system of N atoms in a box which is initially known to be in a definite state, i.e. the position and the velocity of each atom are known. If the atoms obey the laws of classical mechanics, then this information is sufficient to predict the future

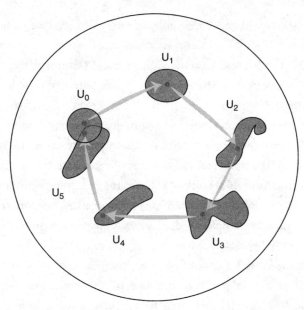

FIG. 29 According to the recurrence theorem of the French mathematician Henri Poincaré, a mechanical system that begins its motion at a point P within region U_0 of the space of states (every point of which is a possible state) must necessarily return to that region at some later time. In this illustration, this happens after five seconds but it may take much longer, in general. After 1 second, the region U_0 has evolved into U_1. One second later, U_1 has evolved into U_2. Yet another second later, U_2 has evolved into U_3, and so on... After n seconds (n = 5 in this example), the region U_n, into which U_0 has evolved, overlaps the initial region U_0. Such a recurrence is a necessary consequence of the fact that the regions U_0, U_1,...U_6,... have identical volumes, even as their shape changes wildly.

motion of each atom. Now comes the surprise. It can be proved with relatively little effort that, no matter how you start, the system will return *after some time* as close as desired to the initial state (Fig. 29). That 'some' is treacherous. It may take a very long time, of course, and the time becomes longer and longer as you make your definition of 'close' more and more stringent. But it must always return, sooner or later. For example, the gas in Fig. 25 will surely fill the box after the partition is lifted, yet our theorem declares as surely that, on some unspecified day in the future, the gas will reassemble in the left half of the box. Clearly, this spells trouble for the Second Law, which flatly denies this possibility.

How does the Second Law manage to stand tall under the weight of the recurrence paradox? The first and most direct answer—the one that was given by Boltzmann and is still found in most physics textbooks—is that the recurrence time is too long to make physical sense. It increases very rapidly with increasing N and gets completely off-scale (i.e. longer than the age of the universe) for systems of moderate size, such as a gas in a box. In the limit of N tending to infinity, which is the limit in which the Second Law sinks in, there are simply no recurrences to talk about. So the Second Law is protected, physically, by a common-sense argument, and mathematically, by a limiting process.

Actually, there is a further layer of protection. In fact, it is a gross oversimplification to say that because of the recurrence theorem a mechanical system is condemned to repeat the same loop again and again. This would be true if the system returned *exactly* to the initial state. But the recurrence theorem only ensures that the system will return 'as close as desired' to the initial state, and there is a huge difference between this and coming back exactly to the initial state. Classical mechanicians were well aware of this, but they probably thought that small differences in the initial conditions would cause correspondingly small differences in the subsequent evolution. Now we know that this is not true. Even for systems of very few particles, the tiniest change in initial conditions gets rapidly amplified as time goes by, leading to completely different futures. This kills our hopes of seeing an exact repeat of what already happened—even in those fortunate cases in which the general pattern is quite easy to predict.

Overwhelming questions

Like a spider at the centre of its web the Second Law is connected by invisible threads to the deepest mysteries of the universe. Free will, for instance. Are we really free to choose a course of action or is our destiny laid out irrevocably from the beginning of time? Newtonian mechanics suggests the second answer. According to this theory, the knowledge of the state of the universe at one time, coupled with the equations of motion, completely and uniquely determines the state of the universe at any future time. Taken seriously, this statement would imply that we are nothing more than automata executing a universal script, deserving neither praise nor blame for our actions. Can this be the truth? Newton himself never believed it. Psychologically, we can hardly doubt our freedom, even under the most extreme circumstances, such as imprisonment, torture, and the threat of death. No matter how perfectly our future is known to God, or to our jailers, we always perceive it as undetermined. And whether it be unnerving or sweetly alluring this uncertainty constitutes the 'wavering pleasure of life, that sometimes is like a fear'.[7]

Does this mean that there is a contradiction between psychology and the laws of classical mechanics? Maybe. But, in any case, the Second Law guarantees that such a contradiction will never be exposed. Indeed, it seems that the only sure way to test free will in a person would be to send her back to the moment in which a certain choice was made. Then she should be given a second chance to choose: if at that point she were able to choose differently from what she did in the past, then she would have convincingly demonstrated the freedom of her will. Unfortunately this return to an exactly identical situation is precisely what the Second Law says will never happen—not on any finite timescale. Restoring the initial conditions means, among other things, erasing the memory of our previous choices and their consequences. So in order to genuinely test free will we ought to erase memory, which is what the Second Law says we can't do—not without creating more memories elsewhere.

The next question is nothing less than the question of the existence of God. Or, more precisely (since nobody can deny the existence of God as a concept in our hearts and minds), the question of the relation between

---

[7] Robert Musil, *The Man without Qualities*, Chapter 21.

God and the World. This relation has been characterized in a great many ways: God the Creator, God the Destroyer, God the Father, God the Mother, and even God the Universe itself. Each role emphasizes a different aspect of the Divinity, and each provides a figurative answer to the question of how much interest, if any, She takes in human matters.

The Second Law suggests yet another model: God the Experimenter. In this model, God sets up the universe as an experimenter would set up an experiment, carefully designed to answer a question. In other words, God does not just create the World, but creates it in a very, very special state, i.e. with an extremely low entropy which can thereafter only increase, generating the Second Law. It is as if, to learn something about the nature of gases, we started an experiment by confining all the molecules of a gas to the tiniest volume in a corner of a huge box, then removed the constraint and watched the gas inexorably fill the box. Like this—but on a cosmic scale—and only in a most abstract sense, for the 'molecules' were not there yet at the time when the experiment began.

According to the modern theory of cosmology, the beginning of the cosmic experiment was a singular event, which marked the birth of space and time. Detractors of the theory called it the *Big Bang* with derisory intent: the name stuck long after the intent was forgotten. Now the problem arises (a problem for us, not for the Creator of course): how could the Big Bang create such a special universe? The problem was to create a very unlikely initial state, such that there would be a lot of room for it to evolve into something interesting—in other words, to be a beginning and not an end. The problem was to find a way to put all the stuff in a corner of the box, when there was no box yet, but only one point.[8]

---

[8] It is tempting—but incorrect—to say that the entropy of the early universe must have been low simply because the universe itself was small. Small and big are relative concepts, and you cannot call the whole universe small when there is nothing outside to compare it to. Regarded as a mechanical system, the universe was as big and complex then as it is now.

The solution devised by the Creator is breathtaking in its simplicity. She spread her stuff (radiation, quantum fields, etc.) uniformly across the early Universe like a pizza maker spreads the sauce and other toppings on the dough (if you find this gastronomic metaphor to be in questionable taste, wait till you read the quote at the end of this chapter). And then She added the key ingredient, gravity, or, at least, the potential for gravity to emerge as the universal attraction that would make everything seek out everything else, as if craving for reconjunction.

That a uniform distribution of the available ingredients would make the early universe absolutely special and unique seems absurd at first. If you or I were to create the world, wouldn't we try to make it more interesting by adding some detail? But this thinking fails to appreciate the role of gravity. A uniform distribution of matter, in the presence of gravity, is indeed a very improbable state. For gravity will, in due time, pull clumps of matter out of this uniformity, and the clumps will ignite stars, and the stars will spin out planets and warm up plants and animals on those planets. All this will happen, in due time, due to the relentless action of gravity. And the amazing thing is that throughout this process of self-organization the entropy keeps rising. That bubbling cauldron we call Universe keeps expanding within its own space of states (an abstract space, not to be confused with ordinary space), filling more and more uniformly all its cracks and crevices. How can this be? Didn't I say that the universe was uniform to begin with?

Thinking about this paradox takes me back to my college days, back to that very special day in which I was nervously pacing up and down the corridor of the Institute of Physics of the University of Pisa, waiting for my turn to face the examination committee of 'Physics 1'. I was nineteen and I had never studied so hard in my life, so I could say in retrospect that I was very well prepared. Nevertheless, as we were being called in alphabetic order, I felt terribly anxious and cursed the unfriendly destiny that had given me a name beginning with a V. At long last my turn came. I sat down, and one of the professors asked me a

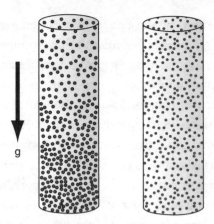

FIG. 30 The effect of gravity on a column of gas in equilibrium. The non-uniform distribution, on the left, is more disordered than the uniform distribution on the right.

question that I still remember vividly because (i) I could not answer it and (ii) it contains the key to the resolution of the low-entropy paradox.

To put the question in perspective I must tell you that, at the time of 'Physics 1', I had been taught that the equilibrium state of a gas of given volume and energy is the state of maximum entropy that is compatible with the assigned values of the volume and the energy. I had also been shown that the molecules of a gas in a box will tend to concentrate near the bottom of the box, due to the effect of gravity. This effect, depicted in Fig. 30, is more pronounced for heavier molecules than for lighter ones—larger for oxygen than for nitrogen—a piece of information that might prove valuable if you ever find yourself locked up in a room with little air: there is always more oxygen near the floor. And here was the question: how is it possible for this state, in which the molecules are denser near the bottom of the box, to be the state of maximum entropy? Remember: entropy is a measure of molecular disorder. So why wouldn't the state of maximum disorder be the one in which the molecules are uniformly distributed throughout the volume of the box?

If you think about it for a moment you will see that the paradox I was asked to resolve is very similar to the paradox of the low entropy of the early universe. In both cases, the problem is to explain how a spatially uniform state can have lower entropy than a spatially non-uniform state. The resolution of the paradox requires quite a bold step into the abstract. That step, however, was far too difficult for me at the time of 'Physics I', and I could not answer the question in spite of the professor's sincere efforts to help me out. Fortunately, another professor came to my rescue, saying that we had not yet studied 'phase space'. Then the two of them spent a couple of minutes explaining the answer to each other, visibly enjoying a discussion of which I understood only that it revolved around a mysterious entity called 'phase space'.

Childhood traumas come back to haunt you in adulthood and now I feel compelled to explain the resolution of this 'paradox', as I understand it some thirty years later. Molecular disorder must be assessed not in ordinary physical space, but in the abstract space whose points represent possible microscopic states of the system. Take a single particle, for example. Its position in ordinary space is not sufficient to determine its state, because two particles with the same position but with different velocities will have completely different futures. Both position and velocity are needed—and, it turns out, sufficient—to specify the state of the particle. So the space of the states (also known as *phase space*) is a kind of doubling-up of ordinary space, in the sense that its 'points' are positions *and* velocities (actually positions and momenta, as we will see in Chapter 8): six dimensions rather than three. And each additional particle carries six additional dimensions, resulting in a total of $6 \times N$ dimensions for the phase space of an N-particle system.

Now back to our paradox. Imagine for a moment that gravity has been turned off. The gas fills the box uniformly, and the average position of a molecule coincides with the centre of the box. Now gravity is turned on. The molecules tend to accumulate near the bottom of the box, causing the average position of a molecule to shift downwards. At first sight, this shift appears to decrease the entropy of the gas, insofar as it creates some degree of spatial order. But spatial order is not the whole story. By adopting a lower stance in the gravitational field the molecules increase their average speed. Remember that an object in a gravitational field slows down as it rises and gains speed as it descends. By the same mechanism, molecules will have a higher average speed when their average position descends to a lower level. But, these molecules with higher average speed are spread over a larger volume in velocity space—hence they generate higher entropy. Furthermore, it can be shown that the gain in entropy due to the higher average speed is larger than the loss of entropy due to partial ordering in space. And this is the reason why the accumulation of molecules near the bottom of the box results in a state of higher entropy than the plain uniform state.

The conclusion of this laborious argument is that the uniform state is not at all the most probable state of a gas when gravity is present. And if—as in the early universe—the state is very uniform and gravity is very strong, then you have an exceptionally improbable state—a state that will spawn all kinds of wonderful and unexpected cataclysms, and God only knows, perhaps, what the end will be.

Indeed the cosmological implications of gravity are of biblical proportions. While on the whole the universe keeps expanding, on a local level one sees matter condense, orderly patterns emerge from uniformity. These processes are driven by gravity. A nebula of rarefied gas condenses as its different parts attract and fall towards each other. As the density increases so does the average speed of the molecules—the temperature—and eventually nuclear reactions ignite, which power the stars. Throughout the process, the entropy has been rising. And now the foundations are laid for the beginning of life. The nuclear reactions that power the stars generate the heavier elements from which living matter will be formed. Out of hot stars cool planets condense—entropy still rising. And round those planets the sky is no longer uniform: a bright hot spot inhabits it and supplies the high-level, low-entropy energy that plants and animals will use and re-radiate into space in degraded form—entropy still rising. Meanwhile the star, under the pull of gravity, maintains its precarious equilibrium, waiting for the day in which, all nuclear fuel spent, it will have to resume its collapse. Then, depending on the star's mass, several scenarios are possible, of which the most intriguing is the one that ends in the formation of a 'black hole'—the cosmic information shredder, which irreversibly absorbs all kinds of matter. Needless to say, the entropy keeps increasing through the formation of a black hole. And it is precisely this feature—I mean the continuing rise of entropy—that makes the collapse of a star so radically different from a time-reversed Big Bang. For at the beginning of the Universe there was order hidden in the gravitational field, which stood, as it were, *aloof from matter* (Roger Penrose's words in *The Road to Reality*). But, at the end of the gravitational collapse, matter and gravity are mixed together in a deadly mess from which nothing can be born.

It is sobering to think that the evolution of life and consciousness may be viewed as part of a larger process of degradation of the primordial order that the Creator built into the universe in invisible form, i.e. not as a spatial pattern, but as abstract geometry. If this view is correct, then the partial order we witness today in the universe comes from the corruption of a higher order that once existed only in the gravitational field. Undoubtedly there is a tension, almost a contradiction, between the idea of a quick dramatic expansion of space (the Big Bang) and that of a universal attractive force (gravity), which made the state resulting from that expansion so improbable and therefore so meaningful.

This game of attraction and repulsion, separation and reunion and overall irretrievable loss, is the theme of Italo Calvino's short story *All At One Point*.[9] Two ageless characters, veteran witnesses of cataclysmic events, bump into each other in an Italian town and start reminiscing about the good old times that just preceded the Big Bang. Even in those days, when they were all pressed together like sardines, there were those who loved each other and those who hated each other, but, above all, there was this one woman—Mrs Ph(i)Nk$_0$—who loved everybody with the love of a mother, and was loved by everybody in return. And then,

---

[9] Italo Calvino, *The Cosmicomics*, 'All at One Point'.

one day, in a thrust of love which soared above the pettiness of the world, she said: 'Boys, if I had some room I'd love to make you noodles!' And at that moment everybody began to imagine the space that would be needed for her round arms to knead the flour, and the space that would accommodate the grain, and the fields, and the mountains, and 'the herds of calves that would give meat for the sauce'. And from that collective imagination arose 'the concept of space, and space proper, and time, and universal gravitation, and the gravitating universe,... and billions of suns, of planets, of fields of grain, and scattered throughout continents Mrs Ph(i)Nk$_o$'s who knead with generous, oily, floury arms, and she, from that time on, lost, and we left to mourn her loss.'

6

Representations

One night I was aimlessly zapping between TV channels, when I was arrested by a scene in which a young man—the image of a mafia gangster—was lying on a bed, his back caressed by an attractive woman. Her gestures were loving, but her words were not. She was urging him to kill his boss. A moment later I knew I was watching *Macbeth*.[1] The players were all different, of course. Macbeth and Banquo had changed into city gangsters. King Duncan was a mafia boss. Lady Macbeth—a ruthless woman embittered by the loss of her child. Undoubtedly, their rough accents had little in common with Shakespeare's high-flown verse. Nevertheless, it was absolutely clear that I was watching *Macbeth*. In spite of all variations, the geometry of the play was clearly recognizable. 'My son, Philly, in this business? Over my dead

---

[1] The title of the movie is *Men of Respect* (William Reilly, 1991).

body!'—said the soon-to-be-murdered 'Banquo' upon hearing the prophecy that his son would become 'Boss of all Bosses'.

As with every drama *Macbeth* is borne out of an abstract core, which can be represented in infinitely many ways. *Macbeth*'s core includes a brave man seduced by a dream of unlimited power; an intimate companion who encourages him to pursue that dream; a bosom friend whom he murders, but whose son he cannot prevent from ascending to power; an arch-enemy who kills him, but whose revenge remains incomplete because Macbeth has no children. As long as these elements are in place *Macbeth* remains *Macbeth*, whether it be enacted by feudal warlords or by corporate executives. The actors change, but the plot is absolute. Shakespeare was well aware of this fact, for in *Julius Caesar*, Act 3, Scene 1, he has Cassio declare, immediately after killing Caesar: 'How many ages hence, shall this our lofty scene be acted over, in states unborn and accents yet unknown!'

The plot of a play is a high-level example of what theoretical physicists call an *invariant*—a reality that does not change in going from one representation to another. Different representations arise naturally as two painters look at the same model from different standpoints. But the model remains one even though the painters see him and paint him quite differently. The model is an invariant. This simple and perhaps obvious fact is so fundamental that the great theoretical physicist Paul Dirac once wrote 'the important things in the world appear to be the invariants of these transformations'.[2] The transformations he was alluding to are changes of representation. In this chapter we will pursue this theme a little further. We will talk about representations, and how they transform into one another, and how they are tied together by an invariant reality that transcends them all.

Geometry and coordinates

People of my generation are not likely to forget the dramatic scene of Stanley Kubrik's movie 2001: *A Space Odyssey*, in which a family of

[2] P. A. M. Dirac, *The Principles of Quantum Mechanics*, preface.

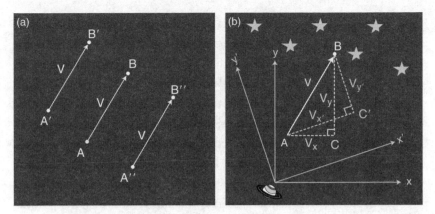

FIG. 31 An abstract vector V in empty space (a), admits different representations in reference frames $x - y$ and $x' - y'$ (b). In the x–y frame, the vector has components $V_x = AC$ and $V_y = CB$. In the $x' - y'$ frame it has components $V_x' = AC'$ and $V_y' = C'B$. The sum of the squares of the components has the same value in the two frames.

prehistoric apes wakes up from a night of fear to find a perfectly polished slab of a mysterious black material stationed in front of their cave. Brushing aside the theory of evolution, the movie suggests that the arrival of the monolith, token of a superior intelligence, is the decisive event that triggers the transformation of the apes into humans.

As one of the apes walks around the mysterious artefact, the view keeps changing. Seen from the front it is wide and flat, but from the side it appears narrow and deep, dangerously sharp. And from below it's like a giant finger pointing to the sun and the moon, which are aligned for the occasion. The front becomes the back and the back becomes the front. Width and length interchange roles. But even so, the ape-man does not doubt for a moment that he is looking at one object, nor do we. Where does this assurance come from? Evidently, there are features that remain constant as the view changes. A square remains a square no matter where you put it, or how you turn it around. Distances and angles appear to change, but we still perceive constant spatial relations between different parts of the object. The walk round the monolith gives our enterprising ancestor a first glimpse of the abstract world of geometry. Classical geometry deals only with such features of objects that do not change when the object is translated or rotated.[3] Latching onto these features enables us to form a steady mental picture of the object. It is the first step towards clear thinking.

Let us take a closer look at the invariants of classical geometry. Consider a segment of straight line running from point A to point B. This oriented segment is the prototype of what in geometry is called a *vector*: let's call it **V**. It takes some effort to visualize **V** as an absolute geometrical object. What I mean is that **V** represents a spatial relationship between points A and B—a relationship that is absolutely independent of the position and orientation of the viewer. In fact, it doesn't matter where A and B are: any two points A′ and B′, which are in the same mutual relation as A and B, define the same vector, the same abstract step in space (see Fig. 31(a)). Now this abstract step 'knows' what its direction is, but the problem is that the information cannot be put into words without introducing a *reference frame*. The simplest way to do this is to draw three oriented lines, known as *coordinate axes*, in three mutually perpendicular directions, such as left–right, back–front, and bottom–top, usually called the x, y, and z axes. For simplicity, only two of them, x and y, are shown in Fig. 31. In the x–y frame, our vector is represented by two numbers V_x and V_y, which specify the magnitude and the direction of the displacement along the x and the y axes respectively. In the present example, V_x is the length of the segment

---

[3] It may be objected that, when viewed at an angle, a circle looks like an ellipse and a square like a rhombus. To deal with such changes we need a more general geometry, in which distances and angles are no longer invariant—a geometry in which the concepts of 'circle' and 'square' have no absolute significance, because they can be transformed into ellipses and rhombuses by some projection. This more general geometry is called 'projective geometry', and forms the mathematical basis of perspective. Even projective geometry, however, has its invariants: two lines that intersect in one projection will intersect in any other projection; four points that lie on a straight line will always lie on a straight line; an abstract quantity called the 'cross-ratio' of four points will have the same value in every projection.

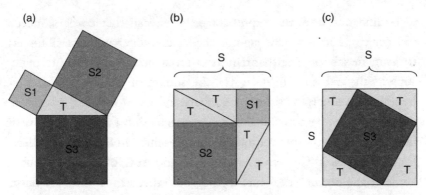

FIG. 32 Pythagoras' theorem (a) asserts that the square S3 on the longer side of the right-angle triangle T equals the sum of the squares on the two shorter sides, S1 and S2. In (b) you see that S1 + S2 equals the area of the big square S minus the area of four triangles T: S − 4T. In (c) you see that S3 is also equal to S − 4T. This proves that S1 + S2 = S3—the fundamental theorem of Euclidean geometry. Notice that the proof relies on the possibility of changing the orientation and position of geometrical shapes without changing their areas.

AC, parallel to the x axis, and V_y is the length of the segment CB, parallel to the y axis. These two numbers are called the *Cartesian components* of **V** in the x–y reference frame.[4] However, we have the freedom to choose a different reference frame—for example the x′–y′ reference frame of Fig. 31. Now the Cartesian components $V_{x'}$ and $V_{y'}$ are the lengths of the segments AC′ and C′B: these are different from V_x and V_y. However, there is no geometric information in $V_{x'}$ and $V_{y'}$ that was not already contained in V_x and V_y: therefore, the former must be expressible in terms of the latter, and vice versa. Furthermore, the length of the step—the distance between A and B—is the same in all reference frames: this implies that an expression for this length in terms of Cartesian components must have the same value in all reference frames.

---

[4] In general, V_x is given a positive sign if AC runs in the same direction as the x axis (as it does in our example of Fig. 31), and a negative sign in the opposite case. Same for V_y with respect to the y axis.

To find out what the expression is, we recall that cornerstone of elementary geometry—Pythagoras' theorem—which we were all forced to learn in school. The theorem asserts that in a right-angle triangle, such as the one shown in Fig. 32, the square of the longer side (the hypotenuse) equals the sum of the squares of the two shorter sides. This theorem can be demonstrated in more than a hundred ways, some very ingenius, but the simple demonstration presented in Fig. 32 is sufficient for our purposes. The two triangles ABC and ABC′ of Fig. 31 are right-angle triangles, and therefore Pythagoras' theorem applies to both. They have the same hypotenuse AB, whose length is just the length of our vector. By applying Pythagoras' theorem to ABC, we see that $AC^2 + CB^2 = AB^2$, and by applying it again to ABC′, we see that $AC'^2 + C'B^2 = AB^2$. Taken together, these two relations imply that

$$V_x^2 + V_y^2 = V_{x'}^2 + V_{y'}^2$$

with both sides of the equation being equal to the square of the length of the vector, AB^2, which has the same value in all reference frames.

The sum of the squares of the components of a step in space is invariant under rotations and translations. This is the fundamental fact of geometry. Well, almost... I should have said, more accurately, that this is the fundamental fact of *Euclidean geometry*—the geometry that was put together by Euclid of Alexandria in the third century BC. Other geometries are possible, in which the fundamental invariant—the expression for the length of a step in terms of its components—is *not* given by Pythagoras' theorem. We will see a spectacular example of non-Euclidean geometry further on in this chapter.

The relation between a vector and its Cartesian components is the simplest illustration of a general pattern. Every abstract geometrical object, be it a point or a plane, a circle or a square, can be represented by numbers and formulas. Recognizing this fact won René Descartes everlasting fame. Although the representation of an object in terms of numbers is highly non-unique—the numbers being strongly dependent on the choice of the reference frame—there are nevertheless some invariant combinations of those numbers, which are independent of that choice. The mathematical expression for the squared length of a vector is just a basic example of such an invariant combination.

Because different representations are equivalent, one might think that it is a big waste of time to study the rules that connect one to the other. Wouldn't it be better if everybody agreed to always use one representation? But it is not so. Different representations are not at all equivalent when it comes to their practical or psychological value. Facts that are hard to see in one representation may become manifest in another. Complex calculations often simplify dramatically when approached from the right angle. And, most importantly, different representations stimulate our imagination in different ways, producing vastly different results. Before seeing *Men of Respect*, I had never noticed how strongly Lady Macbeth's ambition was tied to her being childless.

Space-time

Now let us step out of ordinary space into the wider realm of space *and* time, hereafter abbreviated to *space-time*. Just as ordinary space consists of abstract 'points', which can accommodate infinitesimal things, so space-time has its own 'points', which can accommodate infinitesimal events. And what is an infinitesimal event? Imagine a microscopic explosion taking place at a particular point of space and lasting for just an instant. Every point of space-time can be filled by such an event. And everything we experience in the world can ultimately be described as a coincidence of events—something like: my being here, the light that carries your image being here at the same time. This is the view of space and time that people like Hermann Minkowski and Albert Einstein put forward at the beginning of the twentieth century. According to it, 'space by itself, and time by itself, are doomed to fade away into mere shadows, and only a kind of union of the two will preserve an independent reality'.[5] The laws of physics, like the theorems of geometry, must not depend on the reference frame in which we choose to describe the events.

The main difficulty with this idea is that time and space seem to be made of different stuff. Even disregarding its characteristic feature of flowing only in one direction, time appears to have an absolute character that space lacks. One can, for example, change the orientation of an object in space, so as to make it look sharper or broader. But nothing can be done, at first sight, about the size of a time interval: a minute will be a minute no matter how we turn the clock. Or so it seems...

Einstein decided to re-examine this time-honoured prejudice with a critical eye. It was, by all accounts, a fiercely critical eye. Once, in high school, a teacher told him that he, the teacher, would have been happier if he, Einstein, were not in his class. 'Did I do something wrong?'—asked Einstein. 'No, but the way you look at me is a threat to my authority'.[6]

Einstein's critique of time was no less unsettling. To begin with, he demolished the concept of simultaneity. There is no way, he showed, that two events occurring at different points in space can be said to be simultaneous in an absolute sense. They may be simultaneous in one reference frame, but they will not be simultaneous in another frame that is in motion relative to the first. An immediate consequence of this startling observation is that there is no such thing as an absolute time interval between two events. When you walk around an object, the length and the width get mixed—I mean, what was length becomes width and what was width becomes length. Similarly, in a rotation the x and y components of a vector (Fig. 31) mix into each other. In a completely analogous way time gets mixed with space when we go from one reference frame to a second one that is in motion relative to the first. Uniform rectilinear motion can be viewed as a rotation in

---

[5] H. Minkowski, 'Space and Time', in *The Principle of Relativity* (Dover, 1952).

[6] A. Pais, *Subtle is the Lord* (Oxford University Press, 1982).

FIG. 33 Two people walking past each other in the street have different views of time. For the left walker the cataclysmic explosion that will wipe out life on the Earth is still in the future, but a future as irrevocable as the past. For the right walker it's already in the past, but a past as unfathomable as the future.

space-time, because different times in one frame imply different positions in the other.

Two persons who walk past each other on a sidewalk of New York City have their reference frames tilted, relative to each other, by a very small angle in space-time (Fig. 33). They are looking at the world (if they are looking at all) from two slightly different angles. Their lengths are different. Their time intervals are different. They may even disagree about what is past and what is future. And the only reason why they are unaware of these differences is that the rotation angle is extremely small, something like 0.000 000 1 degrees. One would have to be flying at the fantastic speed of 1 per cent the speed of light (3000 km/s) before the space-time rotation angle approaches the modest value of 1 degree!

You may be wondering how Einstein and his cronies managed to grasp all these counterintuitive facts without hard experimental evidence to support their views (the evidence is available now). The story is fascinating. It all started with a careful study of the electromagnetic theory of light developed by Maxwell in the second half of the nineteenth century. A most important prediction of that theory (also initially unsupported by experiment) was the existence of electromagnetic waves. There is a lot to say about electromagnetic waves and I have reserved an entire chapter, the next, for this subject. The main thing, for our present purposes, is that electromagnetic waves were predicted to propagate in empty space, without need for material support. This fact endowed the calculated speed of those waves with an absolute meaning. If a wave propagates in an external medium (for example, a wave on the surface of water), then its speed of propagation has a clear significance in relation to the medium. But what if the medium is empty space? How can one define a velocity of propagation relative to ... nothing? So, according to Maxwell's theory, the speed of electromagnetic waves in empty space is absolute: it should be the same for everybody no matter how he or she is moving. I will call it simply the speed of light, $c = 299{,}792.458$ km/s, from now on.

If you think about it for a moment you will see how deeply the idea of an absolute speed of light contradicts our intuition. If you were travelling at a speed c towards the source of light wouldn't you expect to see the light coming towards you with a speed 2c? Whereas, if you were taking a ride on a ray of light, then shouldn't the ray have zero speed relative to you? These expectations are in stark contrast with the prediction of Maxwell's theory, according to which the speed of light is always c, irrespective of how you move relative to the source of light. Nevertheless, Einstein believed Maxwell to be right (perhaps because his equations were so elegant) and the absolute speed of light to make perfect sense. He also felt (like Galileo before him)[7], that *all* the laws of physics—not only the laws of electromagnetism—should have the same form for all moving observers, provided they were moving with a constant velocity relative to each other.

Great minds are as free of common prejudices as they are full of their own. Einstein raised his prejudices to the status of postulates, and went on to demonstrate that these postulates were not compatible with

[7] Galileo Galilei, *Dialogue on the Two Systems of the World*, Second Day, 213.

conventional ideas about simultaneity, time, and space. So relativity was not discovered by experimental observation, but by demands of internal consistency of a theory.

Einstein's relativity

Here are the two postulates of the *Theory of Special Relativity* in Einstein's own words:[8]

1. The laws by which the states of physical systems undergo change are not affected, whether these changes of state be referred to the one or the other of two systems of coordinates in uniform translational motion.

2. Any ray of light moves in the 'stationary' system of coordinates with the determined velocity c, whether the ray be emitted by a stationary or by a moving body.[9]

These postulates are exceptionally instructive, particularly because the word relativity never appears in them. There is nothing relative in Einstein's principle of relativity. On the contrary, the first postulate asserts the existence of absolute laws, which have the same form in all reference frames. Space and time intervals will be different to different people, but only insofar as this difference is needed to ensure that something more fundamental (the speed of light, the form of the physical laws) remains the same. In particular, the idea of absolute time must be abandoned if the speed of light is to be invariant. So one might say that Einstein sacrificed absolute time to a higher divinity—a quantity that does not change under space-time rotations (i.e. relative

[8] A. Einstein, *On the Electrodynamics of Moving Bodies*, in *The Principle of Relativity* (Dover, 1952).
[9] Notice that, while the invariance of the speed upon changes of reference frame may be viewed as a consequence of postulate (1), the invariance of the speed of light upon changes in the state of motion of the source is an additional and independent postulate.

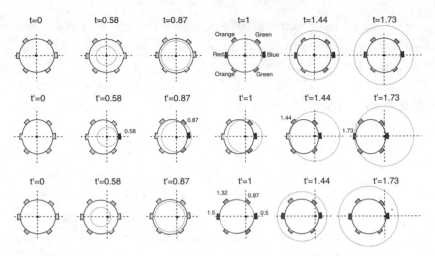

FIG. 34 A flash of light emitted by a point source strikes a set of detectors mounted on a spherical shell, causing them to change colour. The top row shows the progress of the experiment in a reference frame in which the source of light and the detectors are both at rest: all the detectors change colour simultaneously exactly one second after the light leaves the source. The middle row shows the same experiment as seen from a spaceship, which happens to be whizzing by the source at half the speed of light at the time when the light is emitted. *Notice that the light fronts are still circles centred at the origin.* In this reference frame, the rightmost detector changes colour first (0.58 seconds after the emission), followed by the two mid-right detectors (0.87 seconds after emission), the two mid-left detectors (1.44 seconds after emission), and, finally, the leftmost one (1.73 seconds after emission).

The bottom row shows how the experiment would appear from the moving spaceship in the pre-relativistic view of Galileo. The detectors would change colour simultaneously, but light would be travelling at different speeds in different directions.

motions) just as ordinary distance does not change under spatial rotations. Let us see how.

THE RELATIVITY OF SIMULTANEITY

You are standing at the centre of the universe (point 0 in Fig. 34) holding in your hand a camera flash. At time 0, you press the button and a short but powerful burst of light rushes out with speed c in all directions. After a time t the light is uniformly distributed on the surface of a sphere of radius r = c × t which keeps expanding at the speed of light. Now imagine that a huge spherical shell of radius 1 light-second (a light-second is the distance light covers in one second, that is to say, about 300,000 km) has been built around point o (at this point, you probably begin to see why the experiment is unpractical). On this sphere, some sensitive light detectors have been installed, which are predisposed to change from a dull grey to some distinctive colour (red, green, orange, or blue) as soon as they are touched by light. The situation is shown in the top row of Fig. 34. At time t = 1 the light strikes the detectors (the speed of light being the same in all directions) and they all change colour simultaneously—or so you think.

Let us imagine that at time 0 I happen to be whizzing by you at one half the speed of light (another impracticality) going to the right. What will I see? Because the speed of light in my reference frame is still c—just as in your reference frame or in anybody else's—we must assume that the light in my reference frame will still be distributed on the surface of a sphere of radius r = c × t, of which I occupy the centre. But, in relation to me, the detectors are moving: so we can immediately conclude that the detectors on the right, which are moving towards me, will reach the light sphere before the ones on the left, which are moving away from me.[10]

The middle row of Fig. 34 shows the relevant chronology of events in my reference frame. The first detector to change colour, from grey to blue, is the rightmost. This is followed by two detectors turning green, two more turning orange, and, finally, the leftmost detector turning red. So the events that were simultaneous in your reference frame are not at all simultaneous in mine. This conclusion is inescapable as long as we believe that the speed of light is the same in all reference frames.

The bottom row of Fig. 34 shows the hypothetical scenario that would ensue if the speed of light changed according to classical

---

[10] Switching between 'I' and 'you' in these arguments is a fascinating experience. 'You' do not need to be in my reference frame to know what 'I' am going to observe in it. Abstract thought goes beyond reference frames.

intuition. The detectors and the centre of the expanding sphere of light would both be moving together to the left in my reference frame, and the light would still be reaching all detectors at the same time. Then simultaneity would have an absolute meaning, but light would be travelling at different speeds in different directions.

SPACE-TIME ROTATIONS

An analogy with rotations begins to dawn. An ordinary spatial rotation modifies the components V_x and V_y of a step in space in such a way that the square of the length of the step remains unchanged: $V_x^2 + V_y^2 = V_{x'}^2 + V_{y'}^2$. Earlier, I called this invariance 'the fundamental fact of Euclidean geometry'. Is there a similar invariance in space-time geometry?

Yes, there is. To begin with, we must define a 'step' in space-time, i.e. a *space-time vector*. To do this we pick two point-events: one occurs at position A and time t_A; the other occurs at position B and time t_B. The ordinary vector that connects position A to position B is the *spatial component* of our space-time vector. The time interval $t_B - t_A$ is its *time component*. Let us call d the length of the spatial component of AB and t the time interval $t_B - t_A$. The values of d and t change in going from one reference frame to another, just as the two components V_x and V_y of an ordinary vector change in a spatial rotation. But, the combination $t^2 - d^2$ is unchanged, just as $V_x^2 + V_y^2$ is unchanged in an ordinary rotation. In mathematical language we have

$$t^2 - d^2 = t'^2 - d'^2$$

for any two reference frames, provided d and t are expressed in compatible units, e.g. t in seconds and d in light-seconds, or t in years and d in light-years.

To understand why this is so, let us go back to Fig. 34 and consider the events: A—light is emitted at point o, and B—the leftmost detector turns red. In your reference frame (top row) the spatial distance between these two events is d = 1 light second and the time interval is t = 1 second. In my reference frame the distance between the same two events is d′ = 1.73 light seconds and the time interval is t′ = 1.73 seconds. In both frames, the two events are connected by the propagation of a light signal which propagates with velocity c = 1 light-second per second: this is why d = t and d′ = t′. You see that our values of d and t are quite different. Nonetheless, it is obvious that $t^2 - d^2$ and $t'^2 - d'^2$ are equal, since they are both zero!

It turns out that, owing to the uniformity of space and time, the equality of $t^2 - d^2$ and $t'^2 - d'^2$ must hold for *all* values of d and t, if it is to hold in the special case of d = t, which we have just considered. The quantity $t^2 - d^2$ is thus the space-time analogue of the square of the length of a vector in ordinary space. Its invariance is the mathematical expression of the fact that the speed of light is the same in all reference frames.

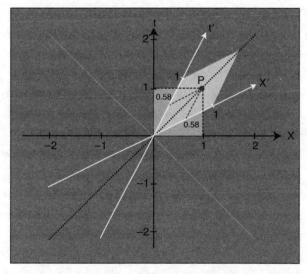

FIG. 35 An event in space-time has different coordinates in two reference frames, which are in uniform motion relative to each other. The relation between the x − t frame and the x′ − t′ frame is such that the area of the diamond spanned by the time and space units on the primed axes equals the area of the square spanned by the corresponding units on the unprimed axes. The trajectory of a light ray is the same in the two reference frames.

To gain further insight, it is a good idea to consider a simplified space-time in which only one spatial dimension exists. Here everything that happens, happens along a line. Events are represented by points in the x—t plane, shown in Fig. 35, where the two axes labelled x and t represent space and time respectively. The coordinate x represents the position of the event along the line, and the coordinate t is the time of the event. The two axes meet at the point x = 0, t = 0, which is called 'the origin' of space-time. A person who stands stationary at the space point x = 0 is described by a point that moves steadily up towards larger and larger values of t, for no one can resist the universal flow of time.[11]

Consider a second person who travels along the x-axis at a fraction of the speed of light (relative to the first), and whizzes through point x = 0 at time t = 0. The question is: how should her axes be oriented, relative to the axes used by the first person? Without going into details, I will only tell you the beautiful answer, which follows directly from the invariance of the speed of light. First, the new axes x' and t' are symmetrically tilted relative to the old ones, in such a way that the diagonal of the x—t plane coincides with the diagonal of the x'—t' plane (Fig. 35). Second, the inclination of the new time axis t' is determined by the speed, v, of the second person relative to the first: in other words, we have x = v × t along the t' axis (this is shown in Fig. 35 for the special case in which v is half the speed of light). Third, the units of space and time on the two sets of axes are chosen in such a way that the area of the diamond spanned by the time and space units on the new axes equals the area of the square spanned by those units on the old axes. The full geometrical construction is shown in Fig. 35. Notice that the tick-marks on the x' and t' axes are more widely spaced than their counterparts on the x and t axes.

[11] Observe how exquisitely relativistic is this 'timeless' view of time as something that is already laid out in its entirety and waiting to be explored. The familiar notion of time flowing by us is replaced by the notion of us flowing through time.

At first sight our transformation of the x–t axes does not look at all like a rotation, since it fails to preserve lengths and angles. But this is only because the invariant quantity is $t^2 - x^2$ rather than the Euclidean quantity $t^2 + x^2$. The geometry of space-time is a non-Euclidean geometry in which the square of the distance of a point from the origin of the x–t plane is given by $t^2 - x^2$ rather than $t^2 + x^2$. That sign difference is the technical reason why the transformation does not look like an ordinary rotation.[12]

Another important property of our transformation is that it cannot be performed for relative velocities larger than c. The largest transformation we can possibly do entails tilting the t′ and x′ axes so that they both lie along the diagonal of the original x–t plane. This largest of transformations corresponds to relative velocity c. We now see that nothing can travel faster than light, if the speed of light is to be invariant.

Space-time diagrams are extremely useful in understanding relativity. We can use them, for example, to analyse the thought experiment described earlier in this section. The event in which the rightmost light detector turns from grey to red is marked as point P on the space-time diagram of Fig. 35. A point is a point and an event is an event regardless of reference frame. However, its space-time coordinates differ. In the x–t frame (your reference frame) this event occurs at x = 1 and t = 1. To find out the coordinates in the x′–t′ frame (my reference frame) I draw, starting from P, two lines parallel to the x′ and t′ axes: the intersections of these lines with the t′ and x′ axes, respectively, determine the new position and the new time of the event.

---

[12] The remaining two directions of space, which we have disregarded so far, turn out to be completely unaffected by a transformation to a reference frame that moves along the x axis. A rotation in the x–t plane does not affect the y and z coordinates any more than a rotation in the x–y plane affects the z coordinate. This is why we could ignore the y and z axes without serious consequences.

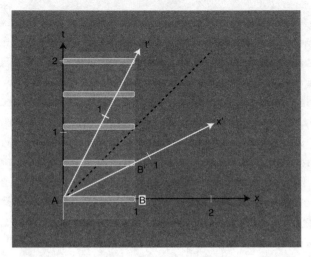

FIG. 36 A stick in motion appears shorter along its directions of motion than the same stick at rest. As the stick changes its orientation in space-time it offers a stationary observer a compressed view of its spatial side (AB′ < AB).

LENGTH CONTRACTION

Examining the second line of Fig. 34 closely, you may notice that the shell that supports the detectors is not exactly spherical in my reference frame: it is compressed along the direction of motion, just as a circle drawn on a sheet of paper looks like an ellipse when you tilt the paper. The strange thing is that there is no way to 'tilt' a sphere in three-dimensional space, for the simple reason that a sphere looks the same in all directions. But, if we add a fourth dimension—time—then it becomes possible to tilt the sphere with respect to the time axis, and we should not be surprised—in fact we should expect—to see a change in its shape!

A 'length contraction' in the direction of motion is another inescapable consequence of the invariance of the speed of light. Consider a simple one-dimensional object—a 1-light-second-long stick lying at rest along your x axis. As time goes by the stick moves up along the t axis as shown in Fig. 36, without changing its position in space. I take a look at the same stick in a reference frame that moves, relative to yours, in the positive x direction, so that my space and time axes are tilted relative to yours, as shown in the figure. What is the length of the stick in my reference frame?

To find out, I must measure the distance between the two ends of the stick *at the same time*. Obviously, it would not do to mark the position of one end now, and the position of the other end at a different time. And here is the rub, because the 'same time' in my reference frame is not the 'same time' in your reference frame. When I pinpoint the two ends of the stick *simultaneously* in my reference frame, I am actually looking at the two space-time points that are labelled A and B' in the space-time diagram of Fig. 36. You see that these two points are nicely aligned along the x' axis— my axis: so they both occur at the same time, $t' = 0$, in my reference frame. But in your reference frame (the x–t frame) these are not simultaneous events: in fact, from your point of view B' is definitely later than A. To determine the length of the stick you do not look at points A and B', but at points A and B, which both occur at time $t = 0$ in your reference frame. In Fig. 36, you see that the distance between A and B along the x axis is 1, while the distance between A and B' along the x' axis is somewhat shorter than 1. So the stick is shorter in my reference frame.

A more sophisticated example of 'length contraction' is shown in Fig. 37.[13] Two spaceships connected by a taut cable start from rest and are programmed to steadily accelerate at the same constant rate, in such a way that they always remain at the same distance and travel with the same velocity. However, as time goes by, the tension in the cable increases, and eventually the cable breaks. How is this possible? Isn't the distance between the two spaceships constant in our reference frame? Yes, but that is not the distance that matters. The proper

---

[13] This apparent paradox is discussed by John Bell in an article entitled 'How to teach special relativity', reprinted in *Speakable and Unspeakable in Quantum Mechanics*, Cambridge University Press, 1987.

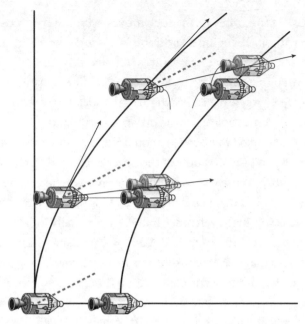

FIG. 37 Two spaceships connected by a cable move with identical speeds and accelerations. As the spaceships accelerate, the proper distance between them increases—even as the apparent distance in our reference frame remains constant. Thus, the tension in the cable increases until the cable breaks.

distance, at a given instant of time, is the one that appears in the reference frame in which the two spaceships and the cable are instantaneously at rest. As explained in the caption of Fig. 37, this proper distance keeps growing as time goes by, causing the cable to strain until it breaks. Equivalently, we can say that the distance between the spaceships is fixed in our reference frame, but the cable becomes shorter and shorter as its speed increases. From either point of view the cable breaks.

TIME DILATION

I had a friend who, upon returning from a vacation, would salute his acquaintances with: 'Welcome back! How have you been?' From his point of view, he had never moved: it was his town, his friends, his family that had moved away and taken a vacation from him.

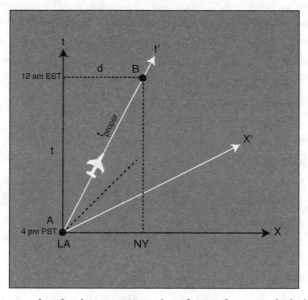

FIG. 38 A trip takes the shortest time in the reference frame in which the traveller is at rest ($x' - t'$ in this figure). The time between arrival and departure in the traveller's reference frame is known as proper time and is given by the space-time version of Pythagoras' theorem: $(t_{proper})^2 = t^2 - d^2$, where t and d are, respectively, the duration of the trip and the distance covered in any other reference frame.

My friend's extreme self-centredness provides a natural way to introduce one of the most important, if not the most important concept of relativity: the concept of *proper time*. Let us say that I travel from Los Angeles to New York at constant speed along a straight line (I'll get rid of these assumptions later). In a reference frame attached to the Earth the trip might look as illustrated in Fig. 38: taking off from Los Angeles, California, at 4 pm Pacific Time (Event A); landing in New York, New York, 5 hours later, i.e. at midnight Eastern Time (Event B). But, in my own reference frame—the x'–t' frame of Fig. 38—arrival and departure happen at the same point, separated only by a time interval t_{proper}, which is dutifully measured by the watch I carry on my wrist. As in a novel narrated in the first person (the most difficult form according to some critics) all that is going on is time. Now, according to the space-time version of Pythagoras' theorem, the square of the space-time length of my trip from A to B, along a straight line, equals the square of the time interval, t_{proper}, *minus* the square of the spatial distance, V in my reference frame. So the square of the space-time distance covered is just $(t_{proper})^2$, and the space-time distance itself is t_{proper}, i.e. the duration of the trip according to my watch.

What about the space-time length of my trip as seen by the person who is anxiously waiting for me in New York? From her point of view I have definitely moved in space as well as in time—the duration of the trip being $t = 5$ hours in her reference frame. But the square of the space-time distance $t^2 - d^2$ in her reference frame has the same value as in my reference frame, namely $(t_{proper})^2$. Two important things follow from this observation.

The first is that the space-time length of a trip is equal to its duration as indicated by the watch that the traveller carries on his wrist. In the jargon of relativity, this is called *proper time*. Notice that in the previous sentence I have surreptitiously changed 'the trip' to 'a trip', implying that the same conclusion applies to curved paths—paths in which the speed and the direction of motion are not constant. This is a natural, but far from trivial assumption. The idea is that the curved path can be split into small segments that can be considered approximately straight. The full space-time length of the path is then obtained by adding up the lengths of the infinitesimal segments.

The second and most amazing fact is that the time duration of the trip for the person who is waiting in the airport (t) is always *longer* than for the person who is travelling (t_{proper}), and not only in a psychological sense: the relation $(t_{proper})^2 = t^2 - d^2$ implies that t_{proper} is always smaller that t.[14]

---

[14] You might ask how to distinguish the person who is travelling from the person who is waiting. After all, each might claim that it is the other who is moving. Answer: the traveller is the person for whom the events 'departure' and 'arrival' occur at the same point, in his or her own reference frame.

TIME AND THE TWINS

Einstein's theory of relativity is known as a theory of narrowly escaped paradoxes. None of the 'paradoxes' is more famous than the *twins paradox*. In the classic version of this parable one of two identical twins is an adventurer who travels around the universe at nearly the speed of light, while the other remains on the Earth—a hard-working stiff in a government job. When, many years later, the adventurer returns to the Earth, he is still young and dashing while his sedentary brother is a tired old man. Notice that the sedentary brother is a space-time traveller too, albeit one who sticks to an uninspiring straight line (Fig. 39). The 'paradox' lies in the fact that the key events 'departure' and 'return' occur at the same point for both twins, putting them in apparently indistinguishable positions: why should one be younger than the other?

A space-time analysis reveals the point where the equivalence of the twins is broken. Consider the paths followed by the twins in space-time from the moment of their separation to the time of their reunion (Fig. 39). Although they start and end together, these paths are markedly different, so, you might say, there is nothing mysterious in their having different space-time lengths. The odometers of two cars driven from New York to Los Angeles along different routes, say via Denver and via Albuquerque, show different mileages at the end of the trip. You wouldn't call that a paradox, would you? But we have seen that the space-time length of a path is just the proper time along that path. Your wristwatch is your odometer in space-time. So you can safely predict that the watches carried by the twins will indicate different times when they are compared at the end of the trip.

The only remaining question is, which watch leads and which lags, or, equivalently, which of the two paths shown in Fig. 39 is 'longer' in space-time? You might be tempted to answer 'the straight path is shorter', because you know that a straight line is the shortest path between two points in space. But that answer fails to take into account the non-Euclidean character of space-time geometry. In fact, one can easily show

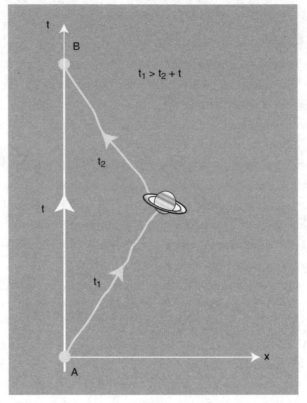

FIG. 39 Proper time is 'length' in space-time. The 'twins paradox' demonstrates that a straight line is not the shortest, but the longest path joining two events in space-time.

that, because of the minus sign in the space-time version of Pythagoras' theorem, the straight line is not the shortest but actually the *longest* path between two points! And this is why the watch carried by the moving twin lags behind the stationary watch.[15]

One might stop here and enjoy some mental rest in the assurance that there is no mystery and no paradox after all. However, the theory of

[15] The effect has been verified in an experiment (Hefele and Keating, 1971) in which two identical clocks are brought around the world on supersonic planes circling the Earth in opposite directions.

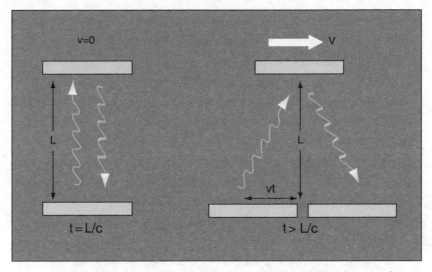

FIG. 40 The *light clock* provides a simple demonstration of the relativity of time. When the clock is stationary, a light ray bounces up and down between two mirrors, each bounce producing a 'tick'. The time between ticks is L/c, where L is the vertical distance between the mirrors and c is the speed of light. When the same clock is set in horizontal motion, the light has a longer way to go from one mirror to the other. Since the speed of light is unchanged, this means that the time interval between ticks is larger. As a result, the clock appears to slow down.

relativity goes much beyond a dry prediction about the behaviour of watches, and asserts that the twin with the earlier time is actually *younger* than the other. Clearly, we are touching on something deeper than the behaviour of watches. It is not just that one watch slows down relative to another, but that every conceivable clock, every device capable to keep track of time, by design or by accident, consciously or unconsciously, must behave exactly in the same way when compared with a copy of itself in a different reference frame. The internal mechanism and constitution of the clock is irrelevant. In fact, the 'clock' that is used in every physics textbook to demonstrate the transformation of time is a highly idealized device, whose only merit is to be analysable more easily and clearly than any realistic clock. It is called 'the light clock' and consists of a single ray of light bouncing up and down between two parallel, perfectly reflecting mirrors, as explained in the caption of Fig. 40.

The importance of the light clock lies in the fact that it gives us a simple way to calculate the rule by which every real clock, indeed every aging human being or beast, must abide. It is the principle of relativity, in the broad sense envisioned by Galileo, that guarantees the consistency of all the clocks. Imagine for a moment that the travelling twin aged at the same rate as his sedentary brother. Then, looking at his light clock, which is lagging behind that of his brother, he would be forced to conclude that the ageing processes in his body were, for some reason, faster than those in his brother's body. But the principle of relativity asserts that all the laws of nature, including the ones that govern the ageing of the human body, are exactly the same on the stationary Earth as in the fast-moving spaceship. So it is impossible for the travelling twin to be older than his clock declares.

Examining the structure of this powerful argument we find it to be very similar to the arguments used in thermodynamics to prove the impossibility of an engine more efficient than the ideal reversible engine (see Chapter 5, page 86). In both cases one compares the behaviour of an ideal device (light clock, reversible engine) with that of a real device (wristwatch, steam engine). Then one makes use of a general principle (principle of relativity, Second Law of Thermodynamics) to establish a relation between the performance of the ideal device and that of its real counterpart. All real clocks must march in lockstep with the light clock; no real engine can outperform the reversible engine.

In the case of relativity, however, the argument leads to a conclusion that transcends the world of physical objects. It would be wrong to think that the relativity of time is due to differences in the inner workings of identical clocks in different reference frames. The whole point of relativity is that the clock's workings are exactly the same in all reference frames. What changes from one frame to another is not the clock, but time itself. You will occasionally hear people say that in Einstein's theory of relativity 'the clock makes the time'. I heartily disagree. Clocks don't make time, they measure it. The transformation of time is a geometric feature of the world, or, more precisely, of the world as we picture it in our minds. Clocks are time's humble bookkeepers.

$E = mc^2$

It should be clear by now that, its name notwithstanding, the theory of relativity is a theory of absolutes, of laws that do not change, of a universal speed of light, of a space-time measure that is the same in all reference frames. The theory, however, allows for multiple descriptions of events: in this consists its relativity. The most famous equation of all time, Einstein's relation between mass and energy, $E = mc^2$, was deduced by requiring consistency between two different descriptions of a light-emission event. It is a wonderful piece of reasoning, which demonstrates dramatically how far—how dangerously far—a bit of clear thinking can go.

Imagine an atom, initially at rest in our reference frame, emitting two identical packets of light in opposite directions, north and south, thereby losing an amount of energy E. Each light packet carries one half of this energy and some momentum along its direction of propagation. However, the total momentum of the two packets is zero, because they propagate in opposite directions. So, if momentum is to be conserved, the atom must remain at rest. The total momentum of our system was zero before the emission of the two packets, and is still zero after the emission. Energy is conserved *and* momentum is conserved. So far so good.

Now what do we see if we look at the same process in a reference frame that moves to the west relative to the original frame? This is shown in Fig. 41. The atom appears to be moving eastwards at a constant speed, while the directions of propagation of the two light packets are slightly angled to the east. This is the phenomenon of the *aberration of light*, a phenomenon known to James Bradley long before the advent of relativity, and used by him to determine quite accurately the speed of light.

Even if you are not a physicist you will perhaps sense something wrong in this picture. But if you are a physicist, a big red light will start

133

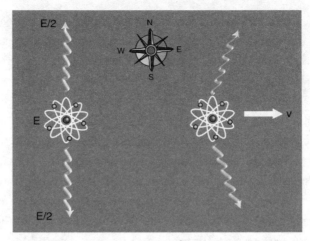

FIG. 41 Conceptual proof of $E = mc^2$. In order to satisfy energy and momentum conservation in different reference frames, the mass of an atom must decrease by E/c^2 when its energy decreases by E.

flashing in your brain when you realize that the total momentum of the system has changed, or so it seems. Just before the emission of light there was only the momentum of the atom, mass times velocity. Immediately after the emission, the momentum of the atom is still the same (according to conventional wisdom neither mass nor velocity have changed), but there is, in addition, a small component of the momentum of the light packets towards the east. Where does this extra momentum come from? Can it be that the law of conservation of momentum is violated in this reference frame? No, that's impossible: the laws of physics are the same in all reference frames. The only possible explanation is that, in spite of appearance, the momentum of the atom has decreased by just the right amount to make up for the extra momentum of the light. But the velocity of the atom has remained constant... Yes, but momentum is mass times velocity. And momentum may decrease even at constant speed, *provided that mass decreases*. The reasoning leads unescapably to one conclusion: that the atom must have lost part of its mass as it was losing part of its energy.

Let E be the total energy carried off by light. Then you don't have to know a lot of physics to be able to calculate the east-bound component of the momentum of the light. It turns out that, according to classical electromagnetic theory, each light packet carries a momentum E/2c along its own direction of propagation. This must be multiplied by the aberration angle v/c (reckoned from the north-south direction) and still doubled (for there are two light packets), in order to give the total east-bound component of the momentum: $(E/c^2) \times v$, where v is the speed of the atom. In order to satisfy the conservation of momentum, the mass of the atom, which moves with speed v, must then decrease by E/c^2.

So far goes the ordinary physicist, but if you are Einstein you will see something deeper, namely that an energy loss must *always* be accompanied by a mass loss, no matter whether it is caused by light emission or by any other mechanism. Indeed the atom could be surrounded by a sophisticated laboratory, whose only purpose is to convert the energy emitted by the atom, by whatever mechanism and in whatever form, into two light packets propagating in opposite directions. The argument could then be applied to the whole lab, regarded as the system that emits the two photons. And again it would be concluded that the mass of the atom must have decreased by E/c^2.

The general conclusion is that mass is essentially the same as the total energy content of a body. If two nuclei collide and stick together losing some mass then the lost mass must resurface as energy: this is the mechanism that powers the stars. And since any change in energy E causes a change in mass $m = E/c^2$ we can say that, in general, $E = mc^2$—the little equation that has been unfairly held responsible for the massacre of Hiroshima, the Cold War, and the nuclear disaster of Chernobyl.

The general theory

Einstein's theory of relativity ushered in the golden age of theoretical physics. After Einstein, and perhaps in spite of Einstein's overall conservative views, physicists felt that they were no longer bound to common sense models of reality, that they had been granted permission to explore a much wider world of possibilities. A door had been opened into the fantastically improbable, and could not be closed again. Quantum mechanics was about to be born. Einstein himself was at work on a general theory of relativity which made the original theory look like children's play. His new idea was that the laws of physics should be formulated in such a way that they have the same form in *every* reference frame, not just in the inertial ones. This meant, in practice, that one should be allowed not only to change the *global* orientation of the reference frame, but also to stretch and compress and distort the space and time coordinates at will: the laws of physics should retain their form.

You will remember from Chapter 3 that Newton's basic law of motion, $\vec{F} = m\vec{a}$ did not have this general validity: it was meant to be used only in inertial frames of reference. You can, of course, *use* Newton's equation in a non-inertial frame. But then you must add to the physical force various ad-hoc forces specific to the reference frame. For instance, in a rotating reference frame, you must include the centrifugal force and the so-called Coriolis force—the force that is responsible for the eastward deviation of falling bodies. These ad-hoc forces seem to be borne out of empty space, rather than arising from natural causes. Their mysterious origin had troubled Newton almost as much as the mysterious origin of the force of gravity.

Einstein realized that the ad-hoc forces, which appear in accelerated reference frames, are a way of encoding different representations of the geometry of space-time. The force of gravity has locally the same character as an ad-hoc force, but with a crucial difference: it cannot be *globally* eliminated by a mere change of reference frame. Rather, it reflects an intrinsic curvature of space-time—a feature that, if present in one frame, must be present in all reference frames.

The geometrical theory of gravity is corroborated by the observation that all objects in a gravitational field move in the same manner irrespective of mass and composition, provided they start at the same position and with the same velocity. This is precisely what one expects of objects moving freely in a curved space-time. According to this view, falling bodies are not falling at all, but simply following the paths of least length in curved space-time. The shape of these paths depends critically on the local geometry of the space under consideration. They are straight lines in a flat Euclidean space, arcs of circle on the surface of a sphere, and they turn out to be parabolas near the surface of a planet like the Earth. This is true regardless of whether the bodies in question are feathers or stones. So the well known Newtonian prediction that a heavy stone and a light feather should fall in exactly the same manner if they are released from the same height with the same initial speed, ceases to be an amazing coincidence and becomes a built-in feature of the physical law.

General relativity is the prototype of what is now known as a *gauge theory*. Electromagnetism was recognized, in retrospect, to be another. Many gauge theories were developed in the twentieth century, and today they form the backbone of theoretical physics. A gauge theory begins with the realization that our descriptions of the world suffer, inevitably, from a degree of arbitrariness. We can change our point of view without changing the facts. And different points of view can create the illusion of a difference, when in actuality there is none. Is there a clear-cut way to distinguish between genuinely different realities and different representations of the same reality? Gauge theories revolve around this central question.

People from different places and times have there own personal measure of things, their own private *gauge* (I might as well say standards and values), which is usually different from their neighbours' gauge. They also have the freedom to change their own gauge at will, without informing their neighbours. This freedom is called 'local gauge symmetry', and the resulting change of gauge is called a 'local gauge transformation'. Clearly, the widespread use of local gauge transformations would make communication between faraway people impossible, unless some common ground is established.

Consider, for example, the arbitrariness in the orientation of reference frames. People from different parts of the Universe will generally disagree on the definition of the cardinal directions, north, east, west and south. Here on the Earth, we might agree to define the north in relation to the North Star, but this definition would be worthless on another planet in a different galaxy. Lacking a common definition of north we and the inhabitants of the other planet could never agree on anything having to do with direction in space.

In order to find a common ground we must develop a protocol for physically transporting a reference frame from our planet to theirs, or vice versa, without changing its orientation. This operation is called *parallel transport*. When two frames are at the same point they can be easily compared and their orientations can be made identical. Then one of the two frames is carefully transported to the faraway location, following the rules of parallel transport. This is done with the help of a 'gauge field', which prescribes how to handle the frame without changing its orientation as we take it from start to destination. By repeatedly applying the protocol one may hope to establish a universal notion of 'north', which everyone, on the Earth and in the remotest corner of the Universe, would find reasonable and acceptable. Similarly, an imperial power might be tempted to parallel-transport its own system of values to every part of the world.

But here we encounter a deep and fundamental difficulty: namely, there is no guarantee that two frames transported from A to B along different paths will coincide in B, even though they were identical in A. Figure 42 shows a Cartesian frame being transported from point 1 to point 3 along two different routes on the surface of a sphere. The two routes produce completely different results. It is even possible to parallel-transport a frame around a closed loop and come back to the starting point with a different orientation—all this while the orientation of the axes purportedly did not change through any small portion of the loop!

When the result of the parallel-transport operation is independent of the path chosen, then and only then is it possible to define an

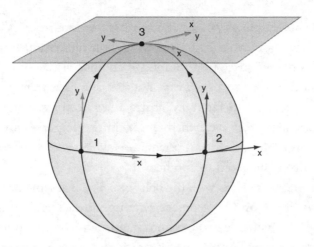

FIG. 42 Transporting vectors on a curved surface is an ambiguous operation. This figure shows the parallel transport of two perpendicular vectors from point 1 to point 3 following two different routes: 1–3 *and* 1–2–3. The pair of vectors that arrives at point 3 following the first route is quite different from the pair of vectors that arrives there following the second route.

absolute orientation in all space. In other words, each owner of a reference frame can perform a rotation that will align his or her frame with everybody else's. If, on the contrary, the result of the parallel-transport operation is a function of the chosen path, then there is no such thing as an absolute orientation, and any attempt to enforce it will inevitably produce inconsistencies. In the first case we say that the underlying space is geometrically flat; in the second case that it is curved. The spherical surface of Fig. 42 is an obvious example of curved space.

The central idea of gauge theory is to interpret force fields as manifestations of some kind of 'curvature'. Curvature, in the most general sense, means that it is impossible to univocally transport a reference frame from a point to another either in ordinary space-time, or in one of the more abstract spaces posited by the quantum theory (see Chapter 9). This impossibility creates a tension between different points of space-time—a tension that cannot be relieved by a simple change of reference frame. In Einstein's theory of general relativity, curvature arises from the impossibility of univocally transporting four-dimensional reference frames from one space-time point to another. The corresponding force field turns out to be gravity. But the idea is extremely general and can be applied to far more abstract connections. For example, electromagnetic forces have their origin in the impossibility of univocally transporting an abstract orientation, which is known technically as 'phase', from one point to another.

Developing these ideas with the help of heavy mathematical formalism, it has been possible to describe rather satisfactorily some of the basic forces of nature, e.g. the electromagnetic force, the 'weak force', and the 'strong force'. Ironically, it is precisely gravity, the most ancient of the forces and the first to be described by a kind of gauge theory, that today is, because of its difficult relation with quantum mechanics, the least satisfactorily described.

A lesson from the Nazis

Few theories in science have been so influential as the theory of relativity, and yet few are so thoroughly misunderstood. Popular imagination has absorbed the idea that 'everything is relative' in the sense that there are no absolute standards of truth. This formula is either enthusiastically embraced or violently rejected, depending on ideological bias. In the USA, 'liberals' are in favour of relativism and what they call cultural diversity, while 'conservatives' oppose it in the name of absolute, usually religious values. These two points of view are not as different as they appear to be. A liberal Christian might regard the Islamic religion as a legitimate alternative to his own. A conservative Christian is likely to think of it as a false one. Neither attitude is relativistic. They both emphasize the difference. A relativist would say that Islam and Christianity are two different representations of the same religious feelings and aspirations. Perhaps also liberals and conservatives are just different representations of the same drive for power.

I started this chapter talking about two different representations of Macbeth and how they shed light on each other. That was an easy case. In most cases the correspondence is invisible, and it takes a great deal of insight to recognize it. The tragedies of Judas and St Peter betraying Jesus have been re-enacted millions of times in human history, under millions of different disguises. And I know of literary works which somehow correspond to each other—because of a certain rhythm, a consonance, a geometry—in ways that I cannot even begin to express.

The great Anglo-American poet T. S. Eliot frames at the very centre of his best known poem, *The Waste Land*, the story of a working girl—a typist—having indifferent sex with a man she does not love. This rather ordinary story acquires the status of a universal tragedy as Tiresias, the blind seer of Thebes, appears to be mysteriously present at the scene: 'And I Tiresias have foresuffered all, enacted on this same divan or bed'. Here the verb 'enact' conveys the crucial message: what is happening is only a representation; the protagonists of the action, the typist and the clerk, enact a larger drama which Tiresias foresaw and foresuffered in the abstract: the murder of purity and innocence.

While T. S. Eliot was writing *The Waste Land* the world was rushing towards great tragedies: the War, the Holocaust, the massacres of Hiroshima and Nagasaki. Eliot genuinely believed that the way to salvation lay in a return to traditional Arian/Christian values and wrote the poem to exalt tradition and condemn the 'squalor' of modern life. But his profound conservatism, his rejection of democracy, his bigotry, his anti-Semitism, made him an unwitting accomplice to the dark movement of destruction that was engulfing Europe.

This brings me to the final point about relativism. Given that we are locked day and night in the prison of our self, that we are keenly aware, above all, of our own needs and our own rights, that we spend most of the time either alone or talking to people who think like us and thus reinforce our views and prejudices—how can we maintain a sound moral judgement? How can we be sure that we are not becoming what the world one day will call, with contempt, with horror, 'the

Nazis'? We should try to imagine how it felt to be a Nazi when to be a Nazi meant to be one of a million respectable, well-adjusted, successful individuals. It felt perfectly normal, as when the space traveller of the physics books crosses the horizon of a black hole, still unaware of his impending fate. But at that point space and time have already exchanged roles and the fall towards the centre has become as inevitable as the progress of time. Just like this, lost in our illusions, lulled into sleep by familiarity with home-spun truths, we become dangerously unaware of who we are and what we are doing.

The Nazis did not know that they were the Nazis. Had they known it, they would have quit immediately. It takes a remarkable intellect to maintain a sense of absolute direction in the night-flight of history. This is the most important lesson from the Nazis and the true meaning of the saying 'all is relative': we could be Nazis without knowing it.

7

The invisible light

A call from the mountain

In the small hours of the night of 11 May 1996, caught in a fierce blizzard on Mt Everest's south summit, at 28,000 feet of altitude, Rob Hall—the leader of an expedition to the world's highest peak—knew that his life was hanging by a thread. Too long had he been waiting for the arrival of a friend, whom he absolutely wanted to reach the top. The friend had succeeded, only to collapse and die on his way down. Now Rob was alone, and nearly out of oxygen. He had a two-way radio, though. The little gadget allowed him to communicate until the very end with his wife in Christchurch, New Zealand. They talked for several hours. 'I am sending all my positive energy your way!'—she said. 'Please don't worry too much.'—he said.[1] Their words were transformed into

---

[1] Source: Jon Krakauer, *Into Thin Air.*

something utterly impersonal, something that could neither be heard nor seen—an electromagnetic field. The field travelled at fantastic speed, undeterred by ice and loneliness, through spaces extending over a quarter of the circumference of the Earth, and transformed itself back into words at the other end. An invisible field linked the distant lovers, as the field of gravity links the Earth and the Moon. A merciful angel, you may think, but actually far more difficult to conceive: no innocent androgynous eyes, no outstretched wings to help our imagination.

When we talk face to face with a person, the physical process of communication is easily pictured. Our speech disturbs the air and the disturbance propagates and eventually reaches the ear of the other person. The whole process takes place within matter. With the help of suitable apparatus we may even succeed in watching the tiny oscillations of the air, which constitute the sound wave. But communicating with electromagnetic waves is quite a different thing. Electromagnetic waves are not the oscillations of a material medium. The presence of air between the source and the receiver is irrelevant at best, and detrimental insofar as it may cause absorption or scattering. Like a stranger who lives in society without belonging to it, an electromagnetic wave travels through matter without material support. And so, the waves upon which our sense of vision is built are, in themselves, invisible. I mean, if you shine light upon light, in a vacuum, you get no reflection, no scattering, no shadow—absolutely nothing. Two beams of light cross each other without revealing anything about each other.[2]

Electromagnetic waves were discovered by a purely theoretical argument, before anyone was able to consciously produce them, let alone use them to convey words and feelings. They were born out

[2] Well, not quite. It turns out that light can interact with light because of a subtle quantum mechanical effect called 'polarization of the vacuum'. More generally, two beams of light can interact with each other through the mediation of matter (think of the extreme case in which a strong laser beam vaporizes the metal screen, which should have otherwise blocked the second beam!) But light does not interact *directly* with itself.

of Maxwell's equations for the electromagnetic field, and Maxwell saw that they could be used to create a theory of light.

As a physicist, I cannot help but wonder what it must have been like to be Maxwell on the verge of this momentous discovery—the sense of power, of thankfulness, of pride, of humility, of not being Maxwell any more, of being touched by immortality—all mixed together in an intellectual and emotional tempest. In this chapter, I'll try to recreate the flavour of that discovery and that tempest. We'll see how two fields, the electric and the magnetic, which were initially nothing more than book-keeping devices, became progressively intertwined, acquired an independent reality, and finally emancipated themselves from matter. But it will not be the real story, for that I do not claim to know. It will be the romanticized story that I learned as a boy in the textbooks, the legend of electromagnetism—no, not the legend, the epic.

Electromagnetic field: an epic in four books

BOOK 1: THE ELECTRIC FIELD

In the beginning was the *electric charge*. This had been known since antiquity, and by Maxwell's times it was an established fact of nature. Matter contains electrically charged particles, which interact among themselves via a force similar to gravity, i.e. with a strength proportional to the inverse of the square of the distance, but enormously more powerful. This is known as the *Coulomb force*. And there is another crucial difference: while gravity is always attractive, the electrical inter-action can be either attractive or repulsive, because the electric charge comes in two flavours, the positive and the negative, and like charges repel, and unlike charges attract. It is for this reason that the tremendous forces of electricity are not immediately evident: the fact that positive and negative charges are almost perfectly balanced implies that the attractive forces cancel the repulsive forces on a sufficiently large scale.

Besides this, the laws of gravity and electricity share another mysteri-ous feature: they let particles act upon each other at a distance, across empty space, without any intermediaries. Popular culture has no problem with this idea: for centuries people have believed that one can summon sickness or death upon a person by casting an 'evil eye'. But nineteenth century scientists were enlightened intellectuals, determined to rid the world of superstition. Perhaps they regarded Newton's acceptance of action at a distance as a residue of medieval mysticism, like his wild alchemic dreams. They believed that physical action requires physical contact. If a body appears to act on another body across empty space, that's a sign that there is an intermediary making contact with both.

Actually, no such intermediary was known at the time, but to feel better about the whole thing they invented a strange way to describe the electric force—a seemingly innocent idea, but one with a great future. The idea was this: every charged particle creates an *electric field*, which completely fills the space. To be sure, the magnitude of this field, denoted by E, decreases as the inverse of the square of the distance d from the source: $E = k/d^2$, where k is a constant equal to about nine billions in modern metric units. So the field is very large at short distance from the source and rapidly decreases with increasing distance, but it is never exactly zero. The charged particle creates the field as a matter of routine, regardless of whether there are other charged particles to be acted upon. It is not as if the particle looked around for other particles, and having spotted one decided to push it or pull on it according to its charge and distance. No: the particle lays out the electric field as a general rule, like a blind impersonal justice, ready to handle every eventuality, yet aware of none in particular. If there is another charged particle in the universe, then the electric field will exert a force upon it. The magnitude and the direction of this force are determined by the magnitude and the direction of the electric field at the position of the particle. The rule is local: only the value of the field at the position of the particle is involved.

The left half of Fig. 43 shows the electric field created by a single unit of positive charge. The right part of the same figure shows the electric field created by two unit charges of opposite signs, $+1$ and -1

Monopole   Dipole

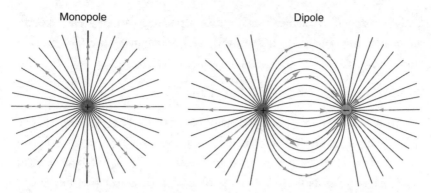

FIG. 43 Left: the electric field created by a single point-like electric charge is shown as a vector at selected points in space. The length of the vector is proportional to the strength of the field and determines the force exerted by the first charge on a second charge located at that point. The solid lines run along the direction of the electric field at each point. Right: same as above for two electric charges of opposite signs.

respectively—an object known as an 'electric dipole'. The field of the dipole is calculated by adding up the fields created by each charge separately, following the rules of vector addition.[3] Observe how the arrows diverge from the positive charge and converge on the negative charge. The rule for computing the force on a charged particle is as follows: the direction of the force is the same as the direction of the field if the charge is positive, and opposite to it if the charge is negative. The magnitude of the force is proportional to the magnitude of the electric field and to the magnitude of the electric charge on the particle. If the charge doubles, the force doubles; if the charge changes sign, the force changes sign.

Up to this point it is not clear that we have gained anything from the introduction of the electric field. The space-filling arrows are an artificial and spectacularly anti-economical way to restate the law of force. Why involve the whole space when we have only two particles at two points? It is not even clear that the bothersome 'action at a distance' has been

[3] Two vectors are added together as if they were two consecutive steps in space. Namely, one must arrange the two arrows so that the second begins where the first ends, and then join the tail of the first arrow to the head of the second.

truly expunged, for it now seems that a particle can create an electric field instantaneously at an arbitrarily large distance. Have we become any wiser?

BOOK 2: THE MAGNETIC FIELD

There is another effect associated with electric charges—or, more accurately, with the motion of electric charges—and that is magnetism. Long before Maxwell began his epic investigations, it had been known that 'magnetic stones' attract pieces of iron and deflect the needle of a compass. In the nineteenth century Hans Christian Oersted had discovered that a wire carrying an electric current has similar virtues. The story goes that he was preparing a demonstration for his students—one of those reassuring demonstrations in which Nature, like a tiger in the circus, is forced to jump through the hoop to the delight of the spectators. The demonstration was, presumably, well-rehearsed, and likely to show the heating of a platinum wire by an electric current. But this time Oersted forgot to remove a compass, which happened to be lying underneath the wire, and when he connected the wire to the battery, the needle suddenly changed direction. It was a small effect, and nearly invisible, but he took notice, and, a few months later, came up with a pamphlet announcing the discovery of a new connection between electricity and magnetism.

Shortly afterwards, André Ampère produced an accurate mathematical description of the force acting between current-carrying wires. He found that two parallel wires attract each other if the currents flow in the same direction, and repel each other if the currents flow in opposite directions. In either case, the magnitude of the force is inversely proportional to the distance between the wires.[4]

---

[4] This seems quite different from the law of the electric force, but it is not. The electric force between two long wires of electric charge (as opposed to two point charges) is indeed proportional to the inverse of the distance between the wires. And the magnetic force between two small current loops (as opposed to two straight wires) is proportional to the inverse of the square of the distance between the loops.

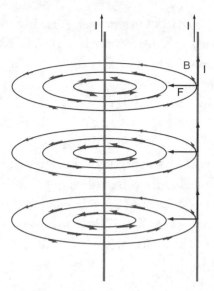

FIG. 44 The magnetic field created by a current-carrying wire (black arrows) exerts a force on a second current-carrying wire. The force is attractive when the two currents are in the same direction (case shown), and repulsive if they are in opposite directions. Only the magnetic field produced by the left wire is shown in this figure.

Here we again encounter the unseemly action at a distance, which nineteenth–century physicists were so determined to ban. The way to do this is completely analogous to what was done for the electric force. One says that the current in a wire creates a magnetic field everywhere in space. It is the magnetic field, not the original current, that acts on the moving charges in another wire. Like the electric field, the magnetic field is a vector, and each point of space has its own vector. The details, however, are quite different.

The first difference is that the magnetic field *circulates* around its source—the electric current—rather than diverging from or converging into point-like charges. So, by following the magnetic field lines, we always end up tracing closed loops, as shown in Fig. 44. As of today, nobody has succeeded in isolating a single magnetic charge—a magnetic monopole—whence magnetic field lines would emanate like electric field lines from an electric charge. By way of metaphor one could say that the current generates the magnetic field like the core of a twister generates the shear winds that whirl around it. The sense of circulation of the field is determined by the direction of the current in the wire: in Fig. 44 the field circulates counterclockwise (looking from above) when the current goes up, but it would circulate clockwise if the current were going down.

The second difference is in the law of force. The magnetic force is perpendicular to both the magnetic field and the current, and proportional to both. You might think that these rules are just a summary and restatement of observed laws of force between current-carrying wires. But they are much more. And new ideas, like fire, have an irrepressible tendency to run wild in all directions.

BOOK 3: ELECTROMAGNETIC INDUCTION

The next gust of wind was Michael Faraday—a humble bookbinder's apprentice, who managed to bootstrap himself into being one of the greatest scientists of all time. One of Faraday's most important discoveries (in physics) was that a time-varying magnetic field generates an electric field.[5] This is known as the principle of electromagnetic induction—the principle that made the power generator possible, and the transformer, and recording the voice of Maria Callas on magnetic tape.

If you move a copper wire in the proximity of a magnet, then the magnetic field will generate an electric current along the wire. This is rather easy to see, because, by moving the copper wire, we also move all the electrically charged particles it contains. These charged particles are of two types: protons, which are heavy and very hard to dislodge from their fixed positions in the atomic cores, and electrons, which are light and spritely and easily accelerated by a force. When you drag a wire across a magnetic field, as shown in part (a) of Fig. 45, the magnetic force, which is perpendicular to the direction of motion of the charges,

---

[5] Faraday also discovered the connection between electricity and chemical reactions, creating ex-novo the branch of chemistry known as *electrochemistry*.

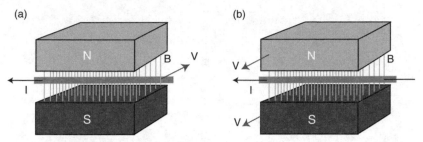

FIG. 45 (a) When a wire is dragged across a magnetic field, the force exerted by the magnetic field on the mobile electric charges within the wire causes a current to flow. (b) Same set up as in (a), but now the wire doesn't move; the magnet does. Again, a current flows.

pushes electrons and protons in opposite directions. The push does not have much effect on the protons, but it drives the electrons steadily along the wire: this is how we get an induced current.[6]

There is really no need to invoke an electric field until you ask yourself the question: what happens if, instead of moving the wire, we move the magnet? Intuitively, we feel that nothing should change: the experiment could be done in empty space, where there would be no way of telling whether the magnet is moving, or the wire. This intuition is absolutely correct. A moving magnet does indeed cause a current to flow along the wire (Fig. 45(b)). But now our explanation of the effect runs into a serious difficulty for, if the wire is not moving, how can there be a magnetic force on the electrons? Remember: the magnetic force is proportional to the *velocity* of the charged particles.

Well, let us turn the question around and ask: what could exert a force on stationary electrons? Answer: an electric field. In some mysterious way, which we do not really understand, the moving magnet must be

[6] The idea that a force makes the electrons move at *constant velocity* seems to contradict the principles of Newtonian mechanics. Shouldn't a force produce an *acceleration*? But in truth the force that drives the current is not the only force that acts on the electrons: there is also a complicated friction-like force generated by the material in which the electrons move. The driving force and the friction force cancel out on average, keeping Newton happy.

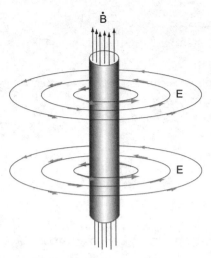

FIG. 46 Faraday's Law of Induction: a time-varying magnetic field generates an electric field in much the same way as an electric current generates a magnetic field (compare with Fig. 44).

creating an electric field, which, in turn, forces the electrons to move along the wire. This is indeed very strange, because the magnet is an electrically neutral object, and we are used to thinking of an electric field as something that is generated by electric charge. But now we see that this cannot be the whole truth: a moving magnet must also be a source of electric field if the principle of relativity is to be satisfied, i.e. if there is no way to distinguish between a wire moving past a stationary magnet and a magnet moving past a stationary wire.

We can be even bolder and speculate that any magnetic field that varies in time—not only the magnetic field produced by a moving magnet—must generate an electric field. This is precisely what Faraday's law of electromagnetic induction asserts. It is an extraordinary law because it connects two abstract concepts, the magnetic field and the electric field, without reference to the material sources that produce them—currents and charges, respectively. The form of this relation is depicted in Fig. 46—and notice how similar it is to the relation that exists between the electric current and the magnetic field. A time-varying magnetic field in a tightly wound coil causes an electric field to circulate around the coil in precisely the same way as the electric current in a straight wire causes a magnetic field to circulate around the wire.

BOOK 4: MAXWELL'S NEW TERM

So this was the state of the electromagnetic theory in the early 1860s, at the time when Maxwell was spending his days pondering the laws of electromagnetism. He was about to make the discovery that would make the great American Civil War 'fade into provincial insignificance' (Feynman's words). It is difficult to fully appreciate Maxwell's discovery without higher mathematics, but I'll try.

Up to this point we have seen that the electric field either emanates from electric charges (Coulomb's law) or circulates around a time-dependent magnetic field (Faraday's law of induction). But the magnetic field is only found to circulate around electric currents (Ampère's law), and does not seem to emanate from any source. Maxwell noticed something strange about this state of affairs. In order for the magnetic field to consistently circulate around currents it is necessary that the current lines never begin or end, but instead close on themselves as magnetic field lines do. But current lines, unlike magnetic field lines, *can* diverge or converge on points where there is depletion or accumulation of charge.

Consider, for example, the situation illustrated in Fig. 47. A sphere of electric charge is collapsing under the action of an increasing external pressure. Clearly, the current lines converge towards the centre of the sphere, and thus fail to close on themselves. Ampère's law for the circulation of the magnetic field likewise fails. Due to spherical symmetry, a magnetic field, if present at all, would have to be oriented along the radius of the sphere and thus be parallel to the current. But, this is in contradiction with Ampère's law, which says the magnetic field must circulate around the current. The simplest way out of the contradiction

B = O?

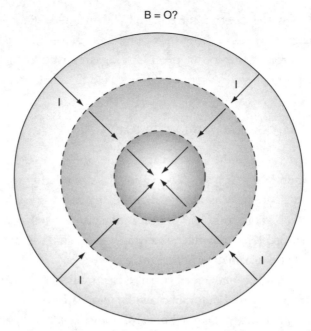

FIG. 47 A collapsing sphere of charge presents us with the paradox of an electric current that produces no magnetic field. The paradox is solved by the inclusion of Maxwell's 'displacement current', which exactly cancels the ordinary current.

would be to say that there is no magnetic field. This is indeed the correct answer, but then we are left with the problem of explaining why the current fails to produce a magnetic field in this case.

Maxwell thought deeply about this problem, and saw the resolution. He saw that the current lines converging towards the centre of the sphere imply an accumulation of electric charge: as the sphere collapses, the electric charge is squeezed into a smaller and smaller region of space. This charge accumulation causes the magnitude of the electric field to grow within the sphere. There is no way of telling what thought flashed at this point in the brain of my fictional Maxwell, but it must have been something like: if a changing magnetic field can generate an electric field (Faraday's law), why can't the reverse happen? Why can't a changing electric field generate a magnetic field? Pure speculation, you

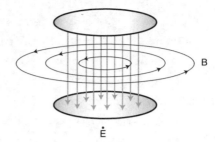

FIG. 48 Maxwell's insight: a
time-varying electric field generates
a magnetic field like a regular electric
current (compare with Fig 44).

see, pure sense of possibility in the spirit of Robert Musil, the great writer who was yet to be born. All Maxwell had to do was to add a new term to the equation for the magnetic field, which soon would be named after him. The new term allows a changing electric field to create a magnetic field (Fig. 48), just as an electric current does. And, to emphasize the similarity between a regular electric current and a changing electric field Maxwell coined the expression '*displacement current*' for the latter.

How does the new term solve the puzzle of the collapsing sphere? Well, I have already said that there is a changing electric field inside the sphere, pointing in the radial direction. If you calculate the displacement current associated with this changing field according to Maxwell's prescription, you find that it beautifully and exactly cancels the ordinary electric current at every point. So, the net current (electric current + displacement current) vanishes and the magnetic field comes out to be zero as we had hoped. Not only the problem of the collapsing sphere, but all the problems connected with the accumulation of electric charge are taken care of in a single shot. And Maxwell was the first to see all this, and he saw that it was good.

Electromagnetic waves

When in the course of human events a new idea makes its appearance in the world, it is at first received with scepticism and suspicion, and oftentimes it is not fully appreciated by its own creators. Maxwell, who undoubtedly had a solid mathematical understanding of the electromagnetic field, does not seem to have attached to his field equations the same importance that we attribute to them. He presented those equations in the final chapter of a lengthy treatise on electricity and magnetism, almost as an afterthought, following scores of detailed examples, which today we would consider illustrations of the general principles. As for his most important prediction—the existence of electromagnetic waves—he struggled to avoid the 'natural' conclusion that the electromagnetic field has an existence independent of matter. To him electric and magnetic fields were still manifestations of an invisible material medium which he believed to fill the whole space. Accordingly, he explained light as electromagnetic waves and electromagnetic waves as mechanical oscillations of that invisible material medium. But the fact is that, through his phenomenal work, the fields had achieved full independence from matter, and there would be no way to push them back into the old subordinate role.

So what are electromagnetic waves? Mathematically, the answer is very simple. They are solutions of the electromagnetic field equations—solutions that propagate in empty space at the staggering speed of 299,792,458 metres per second, also known as 'c'. They are self-sustaining solutions, in the sense that the electric and the magnetic field act as each other's sources and propel each other without the help of matter. The value of the speed of the waves comes out cleanly and easily from the field equations, and is in excellent agreement with the speed of propagation of light, which had already been measured, although not very accurately, in Maxwell's time. This agreement could be interpreted either as an amazing coincidence, or as a clue that light is an electromagnetic wave. Maxwell, of course, opted for the second possibility: 'We can scarcely avoid the inference that light consists of transverse undulations of the same *medium* [emphasis mine] which is the cause of electric and magnetic phenomena.'

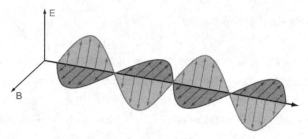

FIG. 49 A snapshot of an electromagnetic wave propagating in a vacuum. The arrows pointing in mutually perpendicular directions represent the electric and magnetic fields respectively. They act as each other's source.

Figure 49 illustrates our present understanding of the propagation of electromagnetic waves in empty space. The electric and the magnetic field oscillate at right angles to each other, and also at right angles to the direction of propagation of the wave. At any instant the three vectors describing the electric field, the magnetic field, and the direction of propagation, form a 'right-handed set of orthogonal axes', meaning that they are related to each other like the thumb, the index finger, and the perpendicular to the palm of your right hand.[7] The changing magnetic field creates a changing electric field (by Faraday's law), and the changing electric field recreates the changing magnetic field via Maxwell's displacement current term. The timing of these coupled oscillations will be maintained, and the dance will go on for ever, if and only if the speed of propagation of the wave is c.

All this explains mathematically how an electromagnetic wave, once established, propels itself in vacuum, but still begs the question of its origin. If we compare the electromagnetic field to a violin string, then we have perhaps explained how the string vibrates, but we have said nothing about the bow that sets the string in motion. This is the radiation process.

[7] The right-handedness of electromagnetic waves in vacuum follows from Maxwell's original equations. Electromagnetic waves in matter are normally right-handed too. However, under carefully controlled conditions they can become left-handed. Left-handed electromagnetic waves have been produced and studied in artificial media during the past few years, and shown to behave in most unusual ways.

Radiation and retardation

Looking back at Fig. 43, we see that the electric field emanating from a stationary electric charge does not go very far: it decreases as the inverse of the square of the distance from the source. What is worse, it doesn't look in any way similar to the electric field of an electromagnetic wave (Fig. 49). The field of the wave oscillates in space, while this one always points away from the charge—a constant reminder of the fact that it emanates from a charge and would not exist without it.

What if the charge were to travel with a constant velocity? This situation is illustrated in Fig. 50. The electric field is slightly compressed in a direction perpendicular to the direction of motion of the charge. Apart from this difference, it still points away from the charge. There is also a magnetic field (not shown in the figure), which is perpendicular to the direction of motion of the charge *and* to the direction of the electric field. Both fields decrease as the inverse square of the distance from the

FIG. 50 The electric field lines of a point charge travelling at constant speed are straight lines emanating from the instantaneous position of the charge. The field intensity decreases as the square of the distance from the charge.

FIG. 51 A snapshot of the electric field lines created by a charged particle which executes small oscillations, up and down, near the centre of the figure. Far from the source, the radiation pattern is clearly visible. Radiation concentrates in a direction perpendicular to the direction along which the particle oscillates. The deep shadows above and below the particle indicate regions from which radiation is nearly absent.

source, which exposes them for what they are: fields tied to matter and incapable of breaking away from it. A true radiation field—a field that is capable of transporting energy and information arbitrarily far from the source—must be transversal, that is to say, perpendicular to the direction of propagation, and must decrease like the inverse of the distance from the source: $1/r$ *not* $1/r^2$.[8] How can we get such a field from a moving charge?

---

[8] This is because the energy flux passing through a sphere of radius r centred at the source is proportional to the *square* of the electric field times the surface area of the sphere, the latter being proportional to r^2. If the field decreases more rapidly than the inverse of r, then the energy flux decreases more rapidly than the inverse of r^2 and the energy flux tends to zero for r tending to infinity.

It turns out that we must *accelerate* the electric charge. This can be done, for example, by driving an alternating electric current in a wire, such as the antenna of Rob Hall's two-way radio. In Fig. 51 I have plotted the electric field created by a single electrically charged particle oscillating up and down near the centre of the picture. The change from Fig. 50 is spectacular. The first striking feature is that the electric field lines (white lines) are perpendicular to the 'line of sight' from the source to the point in which we observe the field. Next, we notice that the field lines evolve in a sinuous pattern, implying that the direction of the field flips periodically as we move away from the source: the period of the spatial oscillation is just the wavelength of the electromagnetic wave. The emission is highly directional, with the electric field being at its strongest in a plane perpendicular to the direction of motion of the source. Finally, and most importantly, the magnitude of the field decreases much more slowly than in Fig. 50—it goes as the inverse of the distance from the particle, which is slow enough to enable the transport of energy to infinity (see footnote 63).

How can such a dramatic change be brought about by a simple acceleration of the electric charge? The answer cuts to the heart of the 'independent existence' of the electromagnetic field. The idea is that a change in the velocity of the particle is not immediately registered by the field at some distant point. It takes some time for the information about the new velocity to travel from the source to the observation point, because information, in general, cannot travel faster than light, and this type of information, in particular, travels precisely at the speed of light. So the further out we are, the earlier on must we look into the history of the source to find the velocity that is presently reflected in the value of the electric field. This *retardation* effect is well known in astronomy. Looking at the remotest galaxies is like looking at the early stages in the history of the Universe, because the electromagnetic field that carries the visual information has taken a long time to reach us. Similarly, the electromagnetic field carrying information about the motion of an accelerated source, takes time to reach the observation point.

Because of retardation, observers located at different points in space will have, at a given time, different views of the oscillating particle. For one observer the particle might be moving up, while for another it is still going down. To a third observer, very far away, the particle might have not yet started its oscillations: for him there is no radiation field whatsoever. At the very least, this explains why the electric field keeps changing direction as one moves away from the source. But how does it explain the long range of the field? How does it explain its direction, essentially at right angle to the line of sight from the particle?

To understand this, we must climb another rung on the ladder of abstraction. I have been talking of information travelling from one place to another as if it was obvious what the information is, but now I must be more precise. In Newtonian mechanics the complete record of the state of motion of a particle contains two items: position and velocity. So it is natural to imagine that a moving particle would send out a signal saying: here I am, and this is my velocity, so that the electromagnetic field can adjust its value accordingly. However, we have just seen that the electromagnetic field depends not only on the position and the velocity of the source, but also on its acceleration. This is a hint that our description of the electromagnetic field is not as efficient as it could be. Perhaps there exists a more fundamental field, which depends only on the position and the velocity of the particles, and from which the electromagnetic field can be derived?

This conjecture turns out to be correct. A careful study of Maxwell's equations reveals that all the information contained in the electric field **E** and magnetic field **B** can be condensed into two simpler fields, the *scalar* potential V (a simple numerical field, without direction) and the *vector* potential **A** together. Taken, Vano **A** constitute the *electromagnetic potential field*.

Now if you thought that **E** and **B** were artificial, here is something even more artificial. The electric and magnetic field are invisible and immaterial, but at least they have an observable effect on matter—they exert forces on electrically charged particles. The electromagnetic potentials do not directly affect matter[9] and are not even uniquely defined, yet they act behind the scenes to produce the electromagnetic fields, which directly affect matter.

What are the electromagnetic potentials good for? For one thing they allow a more economic description of electromagnetism: there are only four potential fields (V and the three components of the vector **A**) as opposed to the six components of the vectors **E** and **B** combined. Once the potentials are known as functions of space and time it is a relatively easy matter to derive the electromagnetic field from the rates of change of the potentials in space and time. The magnetic field is calculated from the rate of variation of each component of the vector potential field in a direction perpendicular to itself. The electric field is a combination of the rate of spatial variation of V and the rate of temporal variation of **A**. For our present purpose, however, the most important property of the electromagnetic potentials is that they are completely determined by the position and the velocity of the charged particle—no need to know the acceleration! It may sound strange, but it is precisely this property that explains why acceleration, combined with retardation, produces a long-range radiation field.

Let us look first at the case of a charged particle that moves with constant velocity. The electromagnetic potentials produced by this charge decrease steadily (as $1/r$) as the distance from the source (r) increases. This is shown by the thick solid curve in Fig. 52. The rate of spatial variation of the potential, and hence the electromagnetic field, is very small: indeed it tends to zero as $1/r^2$ for large r. Retardation is irrelevant because the change in position of the particle is too minute to be noticed (we are at large distances from the particle) and the velocity is, by assumption, constant in time.

But now look what happens when the particle is accelerated. Retardation kicks in because the potentials at different spatial positions reflect

---

[9] Well, most of the time they don't...There are some wonderful quantum mechanical effects in which the potentials act, so to say, in the first person.

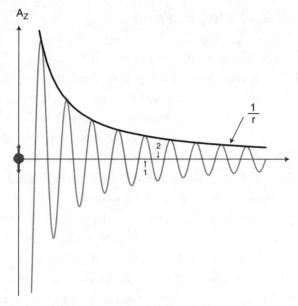

FIG. 52 The electromagnetic potential created by a static or a steadily moving electric charge decreases steadily with increasing distance from the source (black solid line). Now observe the electromagnetic potential created by the same electric charge when it performs oscillations about the origin. This is shown by the grey curve. Notice the sharp variation of the potential between points 1 and 2—as opposed to the mild variation of the black curve between the same points. The

the velocities of the particle at different times. This causes the potential field to change sharply as a function of position. The grey curve in Fig. 52 shows how the vector potential **A** swings from a positive to a negative value when the distance from the particle changes by a very small amount—half the wavelength of the radiation. So, not only the potential itself, but also the rate of change of the potential—that is, the electromagnetic field—is proportional to the inverse of the distance from the source. And this is how the electromagnetic field is able to break away from matter and fly all over the world with its load of energy and information.

The principle of superposition

The discovery of electromagnetic waves opened up a new branch of physics: all the wonderful and multifarious phenomena of optics were reinterpreted in terms of electromagnetic waves roaming in empty space, shaking electric charges, being absorbed and emitted by them. Many years later, after quantum mechanics had popularized the idea that all particles of matter are described by an abstract field—the wave function—the tables were turned around, and physicists came to regard the electromagnetic field as the quantum field associated with 'particles of light', namely, photons. The new and more accurate theory of the interactions between light and matter came to be known as *quantum electrodynamics,* to distinguish it from the classical electrodynamics of Faraday and Maxwell, which I have briefly described.

Both classical and quantum electrodynamics owe much of their predictive power to a very simple feature, which runs deep into the flesh and blood of theoretical physics, and is commonly referred to as the *principle of superposition*. The principle says that the total electromagnetic field generated by several different sources is just the algebraic sum of the fields created by each source individually. Mathematically, this is a consequence of the 'linearity' of the Maxwell equations in vacuum. By this technical expression I mean that only the fields $\mathbf{E}$ and $\mathbf{B}$ appear in the equations, no higher powers such as $\mathbf{E}^2$ or $\mathbf{B}^2$. A linear equation has the property that the sum of two solutions is still a solution: hence the possibility of adding together the fields from different sources. This property implies, among other things, that two waves crossing each other add up together in the region of space in which they coexist, but recover their individual shapes when they separate again.

When light travels through a material medium, things become more complicated. According to the principle of superposition, an electromagnetic wave travelling through a medium such as air or glass is the algebraic sum of two parts. One is the wave that would have existed if the medium had not been present: I will call this the 'wave in vacuum'.

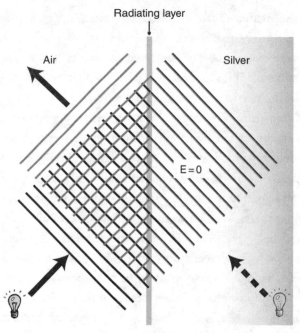

FIG. 53 Reflection of light at a metal surface is explained by the principle of superposition. The incident radiation (light lines) is superimposed to the radiation created by accelerated charges in a surface layer (dark lines). This radiation propagates symmetrically on both sides of the layer. Within the metal, it cancels the incoming radiation. Outside the metal, it is the reflected wave.

The other is the wave that is produced by all the electric charges in the medium as they are shaken by the passage of the wave. This sounds quite complicated, but if you take it seriously and work out the mathematics carefully, you end up with results of amazing simplicity. Adding the waves produced in the medium to the original wave in vacuum produces a total wave that looks in almost every respect like the original, except that it appears to propagate at a different speed than in vacuum.[10] Besides, the wave is gradually attenuated as it progresses

[10] It is believed that every information-carrying feature of the wave (such as the encoding of speech and images) must travel more slowly in the medium than in vacuum. However, some purely mathematical features of the wave, which do not convey information, may travel faster in the medium than in vacuum.

through the medium. The classical explanation for this behaviour is *destructive interference* between the wave generated by the medium and the original wave in vacuum. The energy that is lost to forward propagation is eventually picked up and carried away by a multitude of incoherent motions of the particles of the medium.

One of the triumphs of the electromagnetic theory of light is the explanation of the elementary phenomena of reflection and refraction: they both follow from the principle of superposition. Figure 53 shows how the principle of superposition accounts for the characteristic specular reflection of light by a mirror. A mirror is made of a metal sheet protected by transparent glass. Metals are materials that are essentially impenetrable to electric fields.[11] They contain enormous numbers of electrons, which are not tied to any particular atom and therefore are easily set into motion by an electric field.[12] When one attempts to impose an electric field on a metal, the highly mobile electrons respond dramatically: within a millionth of a billionth of a second they deploy themselves in such a way as to create a counter-field which effectively cancels the applied field within the metal. Thus, an electromagnetic wave of not too high frequency, which impinges on a mirror, is sharply absorbed within a short distance of the surface leaving the bulk of the metal free of electric field.

This miracle—the perfect cancellation of the electromagnetic field within a metal—produces an even bigger miracle, namely the reflection of light and the formation of virtual images. What happens is that the electrons at the surface of the metal, being accelerated, become themselves a source of electromagnetic waves. According to the principle of superposition, the extra field created by these electrons must be added to the field of the incident wave. The apparently difficult sum can be

[11] More accurately, to electric fields that do not vary too rapidly in time. The electric field of a visible light wave, with an oscillation frequency of about 10^{15} cycles per second, is still slow enough to be blocked, but X-rays, with frequencies ten thousand times higher, get through easily.

[12] Non-metals also contain huge numbers of electrons, but those are bound to individual atoms and do not respond as readily to an electric field.

done very easily with the help of two observations. First, the field created by the surface electrons is symmetrically distributed with respect to the surface. So, if we know the field produced by the surface electrons *inside* the metal, then we automatically know the field produced outside: the two fields are indeed 'specular images' of each other. Second, the field that is created by the surface electrons *inside* the metal is just the opposite of the field of the incoming wave since, as we have just seen, the total electric field must vanish inside the metal. Combining these two observations we realize that the wave radiated by the surface electrons back into air is just the opposite of what the incident wave would have been if, upon reaching the mirror's surface, it had sharply changed its direction of propagation according to the familiar rules of specular reflection (i.e. the angle of incidence equals the angle of reflection). Symmetry, superposition, and the perfect absorption of electromagnetic waves in the metal conspire to produce this result of striking simplicity and beauty.

Interference and rays

You will remember that in the first chapter of this book I presented an explanation of the origin of the rainbow. But what an explanation it was! It invoked concepts such as light rays and the laws of reflection and refraction—concepts that are hardly less mysterious than the thing they are supposed to explain. Now, after talking about electromagnetic waves, I would like to return to that explanation and say something about the connection between electromagnetic waves and 'light rays'.

Consider, for example, the phenomenon of refraction. When an electromagnetic wave crosses from one medium into another, its direction of propagation, in general, changes. This is shown in the left panel of Fig. 54, for a wave going from air into water. Notice that the speed of propagation of the wave in water is lower than in air, as you can see from the fact that the wavelength, i.e. the distance travelled by the wave in one cycle of oscillation, is shorter in water than in air. Refraction, like

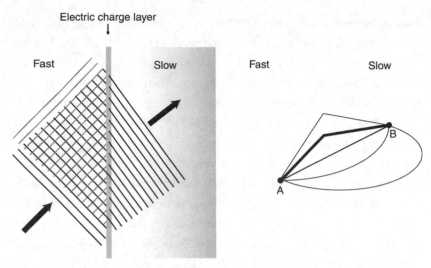

FIG. 54 Left: Bending of electromagnetic wave fronts at the interface between a 'fast' and a 'slow' medium. Right: The same behaviour deduced from the principle of least time for rays: the ray light line goes from A to B in the shortest possible time.

reflection, follows from the principle of superposition. The direction of propagation of a wave in a medium is simply the direction in which the electromagnetic fields produced by all the electric charges of the medium add up constructively: in any other direction destructive interference occurs. One can calculate refraction directly from this idea, but 'light rays' provide an elegant shortcut.

What do we mean when we say that a wave travels along a ray from a point A to a point B? This is a tricky question because an electromagnetic wave is not a particle that goes from A to B, but a field that exists everywhere in space. One might as well say that the wave goes from A to B by an infinite number of different paths, some straight, some crooked, some totally crazy as shown in the right panel of Fig. 54. But among these paths there is one with a special property: it is the fastest one—the one that takes the least time to arrive at B. This path corresponds to the intuitive notion of a light ray going from A to B. In fact, if light were described in terms of particles (as it is in quantum mechanics) this path of least-time would be interpreted as the 'path of the particle', with the understanding that it is not the only possible one, but the most efficient of infinitely many, all of which are simultaneously sampled by the particle.

At a deeper level, it is the principle of superposition that selects the path of least time as the proper definition of light ray. In principle, the electromagnetic field arriving at B from A is the sum of wave-like contributions from all the paths that connect A to B. But these contributions arrive at B at different stages (technically known as 'phases') of their oscillation cycles, depending on how far they travelled, and through which medium. Some contributions will arrive at their maximum (crest of the wave), others at their minimum (trough), others in-between. By and large, these contributions cancel each other out—the trough of one term cancels the crest of another.[13] But there is a tremendously important exception: all the paths that are close to the path of least time arrive at B at essentially the same stage of the oscillation cycle, and thus reinforce each other. Why? Because, by virtue of the principle of the stationary minimum, a small deformation of the path of least time does not significantly change the time of travel, which determines the number of oscillations and hence the 'phase' of the wave that arrives at B. Because of their mutual consistency, paths of almost minimal time constitute an incredibly powerful minority in the general population of paths. While the other paths quarrel with each other, these few act coherently and thus completely control the outcome of the superposition. Their dominance is so pronounced that one usually makes a negligible error by disregarding all the other paths. And this is how one is led to replacing the complicated electromagnetic wave by a single ray of light.

The right panel of Fig. 54 shows the principle of least time in action. The light ray going from A to B changes direction at the interface between air and water. This happens because it saves time to stay a little longer in the fast medium (air) even if that means going a little

---

[13] Strictly speaking, this extensive cancellation occurs only when, as is often the case, the wavelength of the light is the shortest length scale in the game.

FIG. 55 To focus a light beam, a convex lens modifies the path of the rays in such a way that they all take the same time to reach the focal point, where they interfere constructively.

extra distance. You can make practical use of this principle. Suppose you want to build a 'light concentrator'—a device that concentrates the sunlight falling on a disk-shaped area into a single bright spot at the centre of the disk. All you have to do is insert in the path of the wave some light-retarding material of variable thickness, designed in such a way as to ensure that all the paths of least time from the surface of the device arrive at the target spot at the same time. Only when this condition is fulfilled will the paths add up constructively in that one spot. The idea is to slow down the shorter paths more than the longer paths, so that they all arrive at destination at the same time. This is accomplished by making the device thicker in the centre and thinner at the edges. Before you run to the patent office to get credit for this useful invention, I must warn you that it has already been invented: the light concentrator device is nothing but a *convex lens* (Fig. 55).

In Chapter 2, we used the principle of the stationary maximum to explain the characteristic shape of the rainbow, i.e. the fact that nearly all the rays reflected by a spherical water droplet emerge at an angle of 42° relative to the direction of the incoming light. Now we see that the very existence of those rays, the laws of reflection and refraction, and indeed the whole field of physics known as geometric optics, follows from a more abstract application of essentially the same principle. A minimum principle within a minimum principle: Jorge Luis Borges would have to be pleased.

Ether and the void

Abstract ideas can be proved wrong, but rarely disappear from the scene for just that reason. They may suffer a setback, and as a result remain dormant for a long time, but they eventually resurface, sometimes within the very same body of the theory that was responsible for putting them to sleep.

Take, for example, the idea of a perfect emptiness—the void. Newton liked it. In that emptiness he deployed his celestial bodies and his particles. Light was made of particles; gravity acted between particles across the vacuum. When Descartes attempted to fill the vacuum with vortices he, Newton, confuted him.

Two centuries later, Maxwell proposed a theory in which light is an electromagnetic wave. This was as unparticle-like as anything seen until then. But, in order to give a mechanical significance to the electromagnetic field, he found no better way than interpreting it as the elastic displacement of a thin, elusive medium, which filled the universe, and whose only purpose was to support electromagnetic waves. This medium he called the *ether*. The vacuum was gone.

Attempts to observe the ether failed, and Einstein's theory of relativity dealt almost a death blow to the idea. Kids of my generation grew up thinking of the ether as an ancient superstition, one of those fantastic forms of 'dark matter', like the fifth essence, the caloric fluid, flogiston, which had been blasted out of existence by enlightened thought and experiment…only to find out, many years later, that the discredited notion was still alive and well at the forefront of theoretical physics. Indeed, unlike the Newtonian vacuum, which is pure emptiness, the 'vacuum' of modern physics has many features in common with a material medium. Perhaps the reason why we mistake it for nothingness is that our brains are predisposed to perceive only variations against its persistent presence. Ordinary matter and light—so the theory goes—are 'collective excitations' of this medium, just as sound waves are changes in air pressure. Finally, the discovery of a background of cosmic radiation seems to reinstate an absolute frame of reference, that piece of Aristotelian obscurantism that we thought Einstein had eradicated forever.

Every non-trivial idea is, in some sense, right. Throughout history we see the pendulum swinging back and forth between particles and fields, fullness and emptiness. The dry land becomes wet again, the wet land dries up. The opposites propel each other forwards. Will the oscillation ever stop? Will we ever reach an unshakable consensus on 'objective reality'? Personally, I don't think so. As long as there are visionaries, thinkers unwilling to accept what has been passed on to them by tradition and determined to understand things on their own terms, there will always be new ideas, and new angles on old ideas. I am sure that even as we speak there are hundreds of theories sleeping in the limbo of the human imagination, waiting for the kiss of genius to wake up to life. But there is also a lot of resistance from formalized, mindless behaviour; not to mention the curses of weakness and cowardice.

8

The double helix

What I then saw confounded and amazed me. The sweep of the pendulum had increased in extent by nearly a yard. As a natural consequence, its velocity was also much greater. But what mainly disturbed me was the idea that it had perceptibly descended.

With these words the protagonist of Edgar Allan Poe's tale, *The Pit and the Pendulum*, relates his first awareness of the atrocious death that the torturers of the Spanish Inquisition have planned for him. A swinging pendulum, carrying at its lowest end a razor-sharp blade of steel, is being lowered, slowly but inexorably, upon his chest. With desperate lucidity the prisoner studies the motion of the pendulum. He observes details. The velocity is proportional to the amplitude. The blade is descending. Death appears as inescapable as a mechanical law. And yet at the very last moment, in a spectacular swing of fate, the prisoner will be rescued, and the evil Inquisitors will fall into the hands of their enemies.

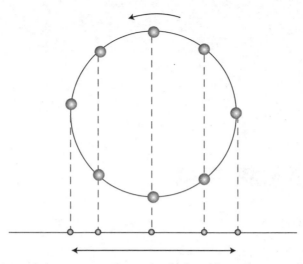

FIG. 56 A simple harmonic motion is the shadow of a uniform circular motion.

Even under less extreme circumstances oscillatory phenomena command a great deal of attention. The great periodicities of day and night, the seasons, the tides, and the phases of the moon, are the mainstays of a clean, well-ordered existence. The mathematicians of antiquity saw a fascinating relation between those cycles and musical sounds. Kepler refers to the celestial cycles as 'Harmony of the World'. This association persists in the modern language, where the simplest type of periodic oscillation is still called a *harmonic motion*.

A harmonic motion is the projection of a uniform circular motion on a straight line. This projection is shown in Fig. 56. As the spherical bead completes a revolution on a vertical circle, its shadow on the ground swings back and forth along a line, its speed being largest at the centre and vanishing at the end points of the trajectory. Harmonic motion is executed by any system that is slightly displaced from equilibrium (for example, a pendulum displaced from the vertical). In an attempt to return to equilibrium, the system overshoots, tries to correct, overshoots again, and so on forever. What makes this motion so special is that it is the most symmetric motion in the universe. It is the only motion in

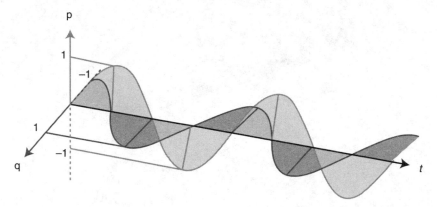

FIG. 57 Evolution of coordinate (q) and momentum (p) in a simple harmonic oscillator.

which the two fundamental variables, position and velocity (or, more accurately, *momentum*, as we will see shortly) play completely equivalent roles.

Figure 57 shows the evolution of the variables q and p (position and momentum respectively) for a particle performing small oscillations about the equilibrium position, which in this figure is at $q = 0$. At the initial time, the particle is displaced from the centre ($q = -1$) and has zero momentum ($p = 0$). Striving to regain equilibrium the particle gains momentum, overshoots, and finds itself on the other side of the equilibrium point ($q = 1$), again with zero momentum. Then the process starts again, this time in the opposite direction. Notice that in each half cycle the momentum increases from zero to a maximum value ($p = 1$ or -1) as the particle passes through the equilibrium point, and then back to zero at the end points.

The next figure (Fig. 58) shows the evolution of q and p for a particle performing a uniform circular motion around a centre—the motion that Aristotle identified as the most perfect and the only one worthy of belonging to celestial bodies. We now know that all linear harmonic oscillations are, in a very precise sense, shadows of this grand celestial motion. Observe how q and p (now vectors in a plane) wind around

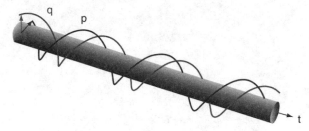

FIG. 58 The double helix of theoretical physics: intertwined time evolutions of q and p in a uniform circular motion.

each other in a circle, keeping a constant angle of 90° with each other. You immediately recognize the pattern of the double helix, better known as the shape of the DNA molecule.

Position and momentum in mechanics are conjugate variables in the sense that they propel each other in time. They are the Yin and the Yang, the double helix of theoretical physics. A uniform motion on a straight line, so simple in appearance, is actually rather asymmetric because the conjugate variables behave in completely different manners: position changes steadily as the particle moves, while momentum remains constant. But in a simple harmonic motion the two conjugate variables are mirror images of each other; they oscillate along intertwined sinusoidal curves, and it is only a matter of convention to say that one is position and the other is momentum.[1]

Figure 49 in Chapter 7 pictured electromagnetic waves as intertwined oscillations of electric and magnetic fields. If you notice a similarity between Fig. 49 and Fig. 57 you are *not* mistaken. The resemblance would have been perfect if in Fig. 49 I had chosen to plot, in lieu of

---

[1] Remarkably, every motion—not only a periodic one—can be represented as a (possibly infinite) sum of harmonic oscillations. Early in the nineteenth century Joseph Fourier realized that the propagation of heat—a distinctly non-periodic process—could be mathematically described as the sum of a large number of simple harmonic oscillations. By so doing, he created the field of mathematics now known as harmonic analysis.

the magnetic field, the spatial rate of variation of the magnetic field. Then you could not have told the difference between the two figures. It is amazing, but the electric and the magnetic fields of an electromagnetic wave can be interpreted as the momentum and the position of an abstract oscillator associated with the wave. And Maxwell's equations for the electromagnetic field are nothing more or less than the mechanical equations of motion for the abstract oscillators of which the electromagnetic field is constituted. This is much more than a casual analogy. Pairs of conjugate variables—of which the fields **E** and **B** are an example—encode the fundamental information about mechanical systems—just as the two strands of the DNA molecule encode the secret of life's myriad forms. In the biological world it is the universal genetic code that, in different combinations, produces the variety of animals and plants. In the world of physics it is the universal equations of motions for pairs of conjugate variables that, driven by different *Hamiltonians* (a word to be explained below), produce the variety of observed motions.

In this chapter I will try to explain the sense in which these obscure-sounding propositions are true. It will be like mountain hiking—sweating and panting until we reach a summit from which we enjoy a truly spectacular vista: a unified view of mechanics and electromagnetism, and, on the horizon, photons and quantum mechanics.

Mechanics and geometry

In Newtonian mechanics, the state of a particle is defined by two variables: position and velocity. These, you might think, are the two conjugate variables that I have been talking about, but it is not quite so. What I am talking about is two variables that stand in a perfectly symmetric relation to each other, similar to the **E** and the **B** of an electromagnetic wave. These variables are called 'canonical coordinates' and 'canonical momenta' and are traditionally indicated by the letters q and p—two letters that have the nice feature of being mirror images of each other.

Canonical variables can be constructed in many ways, corresponding to different but ultimately equivalent descriptions of the system. We shall not dwell on this point. People go to graduate school to learn such tricks. For our purposes, it will be sufficient to think of canonical momentum as the usual Newtonian momentum which equals mass times velocity in the actual motion of a free particle. However, there will be a subtle shift of perspective. In Newtonian mechanics we looked at the momentum, p, as a quantity that could be *derived* from the time-dependent position. Now we are going to look at p as a quantity that, being known at a particular instant of time, allows us to *predict* the position at a slightly later time. A subtle change of perspective, but one with far-reaching consequences. This shift is the essence of the transition from the Newtonian to the Hamiltonian formulation of mechanics. In the Hamiltonian formulation the state of the system is specified by independently assigned values of q and p. So q specifies the position in real space and p the 'position' in momentum space. This combination of real space and momentum space forms a higher dimensional space, known as *phase space* (already encountered in Chapter 5). And it is in phase space that the motion of a mechanical system reveals its deeper significance of geometrical transformation.

At the heart of the formulation is the generator of the time evolution—a function of q and p known as *Hamiltonian* and denoted by the letter H. It is the form of H as a function of q and p that characterizes a particular system. Different systems—for instance a free particle and a harmonic oscillator—have different Hamiltonians. Given the values of q and p at a certain time, the Hamiltonian determines their instantaneous rates of change, which will be denoted by $\dot{q}$ and $\dot{p}$, respectively. From these rates of change we can calculate the values of q and p at an infinitesimal later time. From these values, using again the Hamiltonian, we calculate the rates of change of q and p at that later time, and so on, step after step, until the motion closes on itself—if the motion is periodic—or until the round-off errors become too large or we are too tired. This procedure is summarized in the Hamilton equations of motion, which are exceptionally symmetric and elegant in form:

$$\dot{q} = [q, H] \qquad \dot{p} = [p, H]$$

On the left-hand side $\dot{q}$ and $\dot{p}$ denote the rates of change of q and p in time (Newton's notation). On the right-hand side, the square brackets [q,H] and [p,H] are known as *Poisson brackets,* and have a simple geometric interpretation. I will show you how it works only in the simplest case of just one q and one p, in which case the phase space reduces to the ordinary plane with q on the horizontal axis and p on the vertical axis. But the construction is generalizable to any number of q's and p's.

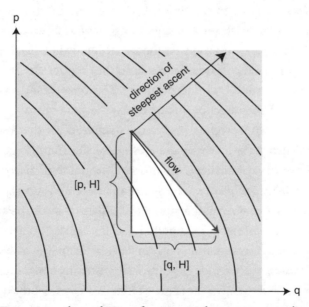

FIG. 59 To compute the evolution of position and momentum we draw the 'level curves' of the Hamiltonian function in the q–p plane. The point (q,p), representative of the state of the system, will flow along these curves with a speed determined by the 'steepness' of the Hamiltonian. The components of the flow along the axes are the Poisson brackets of q and p with the Hamiltonian.

We begin by drawing in the q–p plane the curves along which the Hamiltonian H is constant. These are like contours in the altitude map of a mountainous landscape. In an altitude map all the points on the same contour are at the same altitude. Here it is the value of H that remains constant along the contour. To construct the Poisson brackets [q,H] and [p,H] at a given point in the q–p plane we draw a vector tangent to the contour of constant of H at that particular point. We orient this vector in such a way that, looking along it, the higher value of H (the high side of the mountain) is on the left. We make the length of the vector equal to the 'steepness' of H, i.e. to the rate of increase of H in the direction of its steepest ascent. For reasons that will become clear in a moment, I will call this vector the *flow vector* from now on. The Poisson bracket [q,H] is simply the projection of the flow vector along the q axis. Similarly, the Poisson bracket [p,H] is the projection of the flow vector along the p axis. The construction is shown in Fig. 59.

The content of the Hamilton equations of motion can now be re-stated in geometrical language. According to these equations, the rates of variation of q and p in time are equal to the components of the flow vector along the q and p axes respectively. This means that q and p will always move in the direction indicated by the flow vector (hence the name), with a 'speed' proportional to the magnitude of that vector.

A most interesting scenario emerges when the Hamiltonian of the system does not *explicitly* depend on time. The value of H depends, of course, on the instantaneous values of q and p—but not on time itself. This is a property of systems that are closed and isolated, or at least can be so regarded for reasonably long periods of time. These systems enjoy a symmetry called *translational invariance in time*—a fancy way of saying that they are the same system today as they were yesterday or tomorrow or at any other time. By contrast, open systems can change quickly due to changing environmental conditions. Let us then consider a closed system. Its Hamiltonian, as I said, does not depend on time. In this case, the contours of constant H are also independent of time, and because the flow vector lies along these contours, we see immediately that q and p must move along the curves of constant H. This in turn means that the time evolution of q and p does not change the value of H. Thus H is a constant of motion: it is the 'immobile motor' of Aristotelian philosophy.

It so turns out that this constant of motion is the *energy* of our system. The conservation of energy, which seemed to be a fortunate coincidence in Newtonian mechanics, is 'hard-wired' in the formalism of Hamiltonian mechanics. It is a direct manifestation of translational invariance in time.

Harmonic oscillator

I said earlier that the motion of a harmonic oscillator is the most symmetric motion—far more symmetric than rectilinear motion. I can be more precise now. The Hamiltonian that produces rectilinear motion is $H = ap^2$, where a is a proportionality constant related to the mass of the particle:[2] it contains only p and no q, all Yin and no Yang! But, the Hamiltonian of a particle that performs simple oscillations about an equilibrium point has the form $H = ap^2 + bq^2$, where a and b are positive constants whose numerical value depends on the mass of the particle and the strength of the force that attempts to restore equilibrium. The coordinate and the momentum appear in almost interchangeable roles. In fact, they would be indistinguishable were it not for the different values of the constants a and b. But even this minimal difference can be eliminated by making a transformation that stretches q and contracts p in such a way that the coefficients of q^2 and p^2 become equal. We then arrive at the normal form $H = c(p^2 + q^2)$ where a single proportionality constant, c, determines the frequency of the oscillations.

Look at the right panel of Fig. 60 to appreciate the geometrical significance of this Hamiltonian, and what it implies for the motion of the oscillator. The quantity $p^2 + q^2$ is the square of the distance from the origin of the q–p plane (Pythagoras' theorem again). The contours of

---

[2] For a point particle, *a* is half the inverse of the mass.

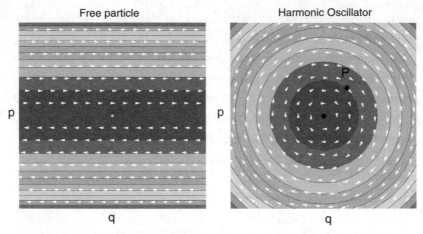

FIG. 60 The flow of a free particle (left) and a harmonic oscillator (right) in the q–p plane.

constant H are therefore circles with centre at the origin (i.e. the point q = o, p = o) and radius proportional to the square root of H. According to the rules stated in the previous section, the point P that represents the state of the oscillator moves along this circle. Furthermore, its motion will be a uniform rotation because all the points around the circle are equivalent, so in particular the magnitude of the flow vector remains constant along the circle. During this rotation the values of the coordinate and the momentum oscillate back and forth with a fixed relationship, meaning that one reaches a maximum when the other vanishes and vice versa (Fig. 57).[3] Then q propels p and p propels q: they are each other's rates of change. Furthermore, the period of rotation is the same for all the circles, that is to say, for all amplitudes of oscillation—a fact first observed by Galileo in the Cathedral of Pisa, when his attention drifted away from the altar to a swinging lamp. This is because the steepness of H is proportional to the radius of the circle, so a point

---

[3] That a circular motion can thus be viewed as the superposition of two oscillations was the basic intuition that led Nikola Tesla to the invention of the electric motor with alternating currents.

moving on a circle twice as large as another moves twice as fast in the q–p plane, and thus completes its trajectory in the same time.

Compared to this elegant dance, the motion of a free particle, shown in the left panel of Fig. 60, appears rudimentary. The contours of constant H are horizontal lines parallel to the q axis. The particle moves to the right or to the left depending on the sign of the momentum, and its speed is proportional to p.

Symmetry and conservation laws

Time evolution under Hamilton's equation of motion is the prototype of a class of geometric transformations that *preserve the form of the equation of motion*. Observe how the time evolution sends a point (q, p) at time t into a different point (q′, p′) at a later time t + dt. We can say that the old variables q and p are transformed into new variables q′ and p′, which continue to obey Hamilton's equations of motion.

This is a general pattern. Not only the Hamiltonian, but any function G of q and p can be viewed as the *generator* of a transformation that preserves the form of the equations of motion. All we have to do is to pretend that G is the Hamiltonian of the system. We obtain new variables q′ and p′ simply by letting the original variables q and p evolve under this fictitious Hamiltonian G for an arbitrary length of time.

Consider, for example the transformation whose generator is momentum itself: G = p. To perform the transformation we pretend that p is the Hamiltonian of the system, and let q and p evolve under it. We follow the rules spelled out in the last two sections. As shown in Fig. 61, the contours of constant p are horizontal lines parallel to the q axis and the corresponding flow is a vector of constant length 1 directed along these lines (this amounts to saying that the Poisson bracket of q and p is 1: [q,p] = 1). Then, Hamilton's equation of motion tells us that the rate of change of q is a constant, 1, meaning that q grows steadily in time. On the other hand, the flow vector has no component along the p axis, which implies that p remains constant. The result of the analysis is that

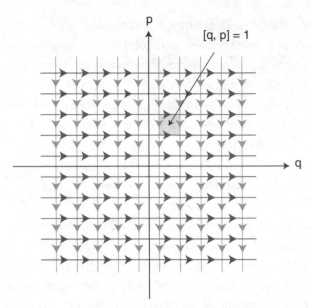

FIG. 61 Geometric construction of the fundamental Poisson bracket [q,p]. The dark and light lines are the level curves of p and q respectively. The dark arrows show how a point (q,p) would flow if the Hamiltonian were p. Similarly, the light arrows represent the flow of a point (q,p) under the Hamiltonian q. These two flows are called the q flow and the p flow respectively. Then [q,p] is the component of the p flow along the q axis (+1), and [p,q] is the component of the q flow along the p axis (−1).

the transformation generated by p is simply a displacement of q with no change in p. Similarly, the transformation generated by q is a displacement of p with no change in q. It is in this very precise sense that I called p the generator of change in q, and vice versa.

Transformations that preserve the form of the equation of motion are really a sophisticated type of rotation in phase space. Recall the fundamental property of rotations: they preserve lengths and angles. Is there a corresponding property for our equation-of-motion-preserving transformations? Yes: and the Poisson bracket is the key to it. A transformation preserves the form of the equation of motion if, and only if, the Poisson bracket of the transformed variables [q′, p′] is equal to that of the original variables [q,p] = 1. For example, the transformation q′ = p, p′ = − q preserves the form of the equation of motion because [q′, p′] = − [p,q] = 1; but the transformation q′ = p, p′ = q, which simply interchanges q and p without the minus sign, does not, in spite of its legitimate appearance.

Symmetry transformations are a special case of equation-of-motion-preserving transformations: a symmetry transformation does more than preserving the form of the equations of motion—it preserves the form of the Hamiltonian itself. From this property a striking result follows: *the generator of a symmetry transformation—let us call it S—is a conserved quantity.* This is the long-anticipated connection (see Chapter 3) between symmetries and conservation laws.

At the root of this connection is a beautiful reciprocity theorem: the fact that the Hamiltonian H is invariant under the transformation generated by S, implies that S is invariant under the transformation generated by H: the two properties have the same mathematical expression. But the transformation generated by H is the time evolution of the system, hence S is a constant of the motion. In this manner one can prove, for example, that symmetry under displacement of the coordinate q implies conservation of the momentum p; this is the previously announced connection between conservation of momentum and uniformity of space.

Oscillators of light

In Chapter 7 we saw that the electric field **E** and the magnetic field **B** in an electromagnetic wave oscillate at right angles to each other in a plane perpendicular to the direction of propagation of the wave. Although Fig. 49 was meant to display the spatial dependence of the fields at a given time, it may as well be viewed as showing the time variation of the fields at a given point. The time evolution of the electric and magnetic field is controlled by Maxwell's equations: the rate of change of **E** in time is proportional to the rate of variation of **B** in space; conversely, the rate of change of **B** in time is proportional to the rate of change of **E** in space.

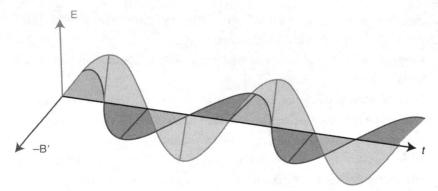

FIG. 62 The electric field and the rate of spatial variation of the magnetic field are the q and the p of a plane electromagnetic wave. Their graph is indistinguishable from that of q and p in a harmonic oscillator (see Fig 57).

The connection between **E** and **B** reminds us of the connection that exists between the 'coordinate' q and the 'canonical momentum' p of a particle executing harmonic motion. In that case the equation of motion says that the rate of change of q is proportional to p, and the rate of change of p is proportional to − q. Could it be that **E** and **B** are just the coordinate and the momentum of an abstract oscillator—actually, one of an infinite number of oscillators that invisibly fill space?

Well, almost. If you compare carefully Fig. 49 and Fig. 57 you will notice that **E** and **B** oscillate in phase (peak matches peak), while q and p are a quarter period out of phase (the peaks in one match the zeroes in the other). Does this indicate a breakdown of our analogy? Not really. Read again the above statement of Maxwell's equation of motion for **E** and **B** and compare it with the corresponding statement of the equations of motion for q and p. Notice that if the electric field is identified as q, then the role of p is not played by the magnetic field, but by the rate of variation of the magnetic field in space. This spatial rate of variation—let us call it **B′**—looks very much like **B** itself, except that it is shifted by a quarter period of the wave. In Fig. 62, I have redrawn Fig. 49 to show **E** and **B′** versus time. The new graph looks exactly like that for q and p in Fig. 57, showing that the electromagnetic wave is indeed a harmonic oscillator of a kind.

What about a general electromagnetic field? According to the principle of superposition, any field can be expressed as a superposition of waves travelling in different directions and with different frequencies. Each wave is an oscillator. The electromagnetic field is therefore equivalent to an infinite assembly of independent oscillators, one for each wave, each oscillator being characterized by a direction of propagation, a frequency, and a polarization. Electrically charged particles, such as electrons and protons, act on the electromagnetic field by pushing or pulling on its invisible oscillators. The oscillators' response is analogous to that of a spring that is extended or compressed by external forces. All processes of radiation, scattering of light, diffraction, reflection, and refraction can be described in this language. The electromagnetic field exists, and always existed, independently of electric charges.

I cannot resist the temptation of ending this chapter with a few words about the photon—the quantum of light. Although this concept properly belongs to quantum mechanics, its roots dig deep into the soil of classical mechanics. In quantum mechanics one still has q's and p's, but they are no longer represented by numbers: they are *operators*. An operator, as the word suggests, is something that performs an operation on the system, such as a rotation, a translation, a reflection: in general, it transforms one state into another.

The idea that physical quantities are operators seems far fetched at first, but on closer inspection we see that it is already implicit in the structure of classical mechanics. After all, we have seen that every dynamical variable can be interpreted as the generator of a transformation. Well, transforming things is precisely the job of an operator. But in classical mechanics the dynamical variables remain conceptually distinct from the transformations associated with them; in quantum mechanics the distinction disappears. In quantum mechanics dynamical variables are considered operators *tout court*: 'momentum', for instance, is the operator that changes the position; 'position' is the operator that changes the momentum,

'Hamiltonian' is the operator that propels all dynamical variables in time, and so on. The fundamental difference between classical and quantum mechanics arises from the fact that the outcome of the application of two operators depends, often dramatically, on the order in which the operators are applied: the operation AB, in which B is applied first and then A, differs from BA in which A is applied first and then B. Anyone who has ever played with a Rubik's cube is aware of this. So the most important thing in quantum mechanics—the thing that ultimately determines the structure of the theory and its predictions—is the *commutator*, defined, for any two operators A and B, as the difference between AB and BA.

The commutator plays such an important role in quantum mechanics that a compact notation has been introduced to represent it: one writes [A,B] as a shorthand for AB-BA. This choice of notation—the same that we used for the Poisson bracket—is not coincidental, for the commutator is the quantum mechanical analogue of the Poisson bracket. It is widely believed that this analogy was first perceived by one of our heroes, Paul Dirac, during a long walk on a Sunday afternoon. He did not shout 'Eureka!', but checked his excitement through a restless night until Monday morning, when the library opened and he could verify the correctness of his intuition.

Most important in this scheme is the commutator between q and p, which is, in a sense, the mother of all commutators. Drawing inspiration from the classical Poisson bracket [q,p] = 1, Dirac posited the basic 'quantum condition' as [q,p] = ih/2π,[4] where h is a universal constant, known as the Planck's constant, and i is the imaginary unit, defined by mathematicians as the square root of -1.[5] Amazingly, h turns out to be the same constant that controls the energy of photons.

---

[4] The 2π in this expression is introduced for mathematical convenience and has no deep significance. The combination h/2π is usually denoted by $\hbar$ (h-bar). It is said that hard-core theoretical physicists cannot write their h's without the bar.

[5] It is far from clear at this point where this mysterious imaginary unit i comes from, but its presence is inescapable if one wants q and p to have *real* values! More about this in Chapter 8.

To understand why it is so, we must first appreciate a most important consequence of the fact that q times p differs from p times q, namely, the quantization of the energy levels of a harmonic oscillator. In classical mechanics the energy of an oscillator can have any positive value: we can say that all the constant energy contours in Fig. 60 are admissible trajectories. The non-commutativity of q and p, with commutator proportional to h, restricts the allowed energies to such an extent that the only admissible values are of the form $(n + 1/2)h\nu$, where ν is the frequency of the oscillator and n is a non-negative integer (n = 0, 1, 2 ...). Alternatively, we may say that only certain special contours in the q-p plane of Fig. 60 are allowed trajectories for the particle; namely, the ones that enclose an area $(n + 1/2)h/2\pi$. This restriction implies that the energy of a harmonic oscillator of frequency ν can only change by integer multiples of $h\nu$. Energy gains or losses must come in packets of magnitude $h\nu$, and a packet cannot be split into smaller parts.

Now comes the real test of our faith. Remember that the electromagnetic field of a plane wave is equivalent to a harmonic oscillator, with the electric field and the spatial gradient of the magnetic field playing the roles of q and p. Are we bold enough to believe that the quantum conditions derived for the mechanical oscillator will also hold for the abstract oscillator defined in terms of **E** and **B**? If so, we must have the courage to *predict* that the energy of an electromagnetic wave of frequency ν can only change in multiples of $h\nu$, that there must exist a quantum of electromagnetic energy $h\nu$, presumably carried by a single 'particle of light', or 'photon', and that any definite-energy state of the wave is characterized by an integer n—the number of photons that are present in that state. If these ideas are correct, then the energy of a photon must be tied to the frequency by the relation $E = h\nu$, where h is the proportionality constant that appears in the commutator of q and p.

Does this sound like a long shot at a theory? Undoubtedly. And, I am sorry to say, it is not the way photons were discovered. In fact, by the time these ideas crystallized, in the mid 1920s, there was already ample evidence that light is produced and absorbed in lumps of energy $E = h\nu$.

Already, Planck had introduced his constant, $h = 6.63 \times 10^{-34}$ joule-seconds, defining it as the constant of proportionality between the energy of the hypothesized photon and the frequency of the light. Already, Einstein had resorted to photons to explain the photoelectric effect. So the creators of quantum mechanics already knew that $E = h\nu$ and set up the commutator of q and p so as to recover the known result. What was not, and could not be, known at the time was that this 'canonical structure', with this value of h, would predict the excitation spectra of systems far more complex than the simple harmonic oscillator or the hydrogen atom: many-electron atoms, molecules, liquids, and solids. It did it with such a spectacular accuracy that even those who disliked quantum mechanics on philosophical grounds had to bow to its predictive power. Sceptics were silenced, and not a single experiment performed ever since has shown quantum mechanics to be wrong.

Today, quantum mechanics undisputedly rules our understanding of the microscopic world, even though many physicists suspect that it is not yet the last word on the subject. Unquestionably, it is an intellectual Taj Mahal, but the ultimate beauty of the Taj Mahal is that it may inspire new Taj Mahals. In this chapter, I have tried to show how this incredible theory is grounded in an abstract geometrical structure, which emerged from a careful study of classical mechanics, starting with simple systems like the pendulum. This structure is so fundamental that it embraces both particles and fields, links the ethereal world of light to the leaden world of matter, and will likely remain an essential component of every future theory.

Maxwell wanted to explain electromagnetic waves as the vibrations of a medium obeying the laws of mechanics. He was absolutely right, but the medium of his dreams was the electromagnetic field itself—an infinite collection of invisible harmonic oscillators.

Quantum mechanics: the triumph of the abstract

Always, my dear Sir, I wish I could see the things
as they are before they show themselves to me.
They must be so beautiful and calm.

Franz Kafka, *Conversation with the supplicant*

In the cockpit

At the end of the second year of college I was impatient to get started with quantum mechanics. During that year I had attended an excellent course on classical mechanics, and I had been told that Hamiltonians, phase space transformations, Poisson brackets, were all steps in preparation of the giant leap: quantum mechanics. I felt like a young pilot entering for the first time the cockpit of a modern jet, admiring the instruments on the dashboard. 'Gone are the days—I thought—when you were learning to fly your little propeller plane, which you could almost steer with the weight of your body, without

191

losing sight of your friends on the ground. From now on, you will be up in the blue empty sky where there is nothing to see and to trust but instruments, instruments that will tell you where you are and where you are going, even though all around be night and mist. And you'd better learn to make the best possible use of those instruments, for your survival depends on it. The mathematical formalism of wave functions and operators will keep you afloat above a microscopic world that is too small to see, but not to imagine.'

Quantum mechanics was invented early in the twentieth century by a group of young men who wanted to resolve some glaring contradictions in the theory of atomic phenomena. The atomic theory of matter was by then well established—the existence of atoms widely accepted. A basic problem remained, and that was the problem of explaining the stability of atoms and the constancy of their properties. A hydrogen atom is, for all practical purposes, eternal. Furthermore, any two hydrogen atoms are absolutely identical: they have the same 'radius', the same electromagnetic spectrum, the same everything. Classical physics did much worse than failing to explain these facts: it predicted a diametrically opposite behaviour.

According to Newton's mechanics and Maxwell's electrodynamics the hydrogen atom, pictured as a miniature sun-Earth system with the proton as the sun and the electron as the Earth, should have collapsed in a flash of light within a billionth of a millionth of a second. Furthermore, there was no apparent reason for the constancy of the atomic properties; in theory, one should have expected to find all kinds of hydrogen atoms, large and small, with the electron orbiting near to, or far from, the nucleus. But this was not the case: like our imagination, Nature selects only one of many possibilities—one that persists indefinitely in time.

It was so that they could understand atoms, and leave their mark in the world, that these young men set out to develop a new mechanics. The solution they came up with had no precedents in the history of science and changed forever the way physics was done. They invented a new pattern of behaviour, strange but coherent, which was neither that of particles nor that of waves, nor that of any object known until then. They gave precise mathematical rules for predicting the outcomes of experiments, and these rules are so subtle, and at the same time so wonderfully artificial, that scientists and philosophers are still arguing about their interpretation. Paraphrasing Rainer Maria Rilke, the great German poet of that generation, one might say that we began to understand the microscopic world only when we no longer understood it.

The mystery

In spite of much recent progress in quantum pedagogy, one of the best introductions to quantum behaviour remains Feynman's description of the 'two-slits experiment'.[1] When Feynman wrote his lecture, the two-slits experiment for electrons was still in the realm of thought, even though actual observations had convinced everybody that, if the experiment could be done, its outcome would be what Feynman described.

By now the two-slit experiment has been done, and quantum mechanics has passed the test with flying colours. Furthermore, it has been possible to observe quantum behaviour of particles much larger than electrons, e.g. whole atoms, confirming that this is, in all likelihood, the universal behaviour of matter.

So let us consider a beam of 'particles'—say electrons—falling on a screen in which two narrow slits, 1 and 2, have been cut. Behind the screen is a movable detector which registers the arrival of a particle by producing an electric pulse which is subsequently amplified to a 'click' in a loudspeaker. By counting the number of clicks we count the number of electrons that arrive every second at a given position behind the screen. The proposed experiment is done first with slit 1 open and slit 2 closed; then with slit 2 open and slit 1 closed; and finally with both slits open. The results of the three experiments are shown in Fig. 63.

---

[1] R. P. Feynman in *The Feynman Lectures On Physics*, Vol I, Chapter 37.

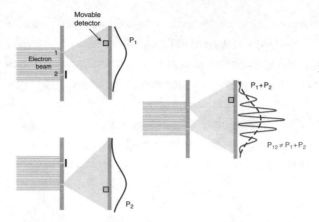

FIG. 63 The two-slit experiment illustrated.

In the first two experiments one finds a broad distribution of particles on the back screen, with the largest number of electrons striking the screen just behind the open slit. There is already a small surprise here. It is reasonable that the largest concentration of particles should be found behind the open slit, but why is the distribution so broad? A 'fan-out' effect is part of the established behaviour of classical waves passing through an aperture narrower than their own wavelength. Whether they are electromagnetic waves, sound waves, or waves on the surface of water, they invariably spread out behind the aperture—a phenomenon known as diffraction. So the spreading of the particle beam behind the aperture already suggests that our 'particles' may have some kind of *waviness* associated with them.

The big surprise, however, comes when both slits are open. Surely any person of a sound mind who has not yet been exposed to the mysteries of quantum mechanics would reason that an electron that reaches the back screen may have gone either through slit 1 or through slit 2, but not through both of them at the same time. This reasoning leads to the expectation that the distribution observed with both slits open must be the simple sum of the distributions observed with either slit open. The expected distribution is shown as a dashed line in

Fig. 63. What is observed, instead, is a striking pattern of maxima and minima—shown as a solid line in Fig. 63. There are prominent peaks, in which the number of electrons is *twice* the expected value. And, between the peaks, there are 'dead spots', which the electrons don't seem to wish to strike.

This pattern is absolutely inexplicable for particles that obey the conventional either/or logic, i.e. particles that either go through slit 1 or through slit 2. For waves, however, it is absolutely natural. Back in 1803, Thomas Young had shown that waves passing simultaneously through two apertures produce an interference pattern that looks exactly like the solid line of Fig. 63. What happens is that the waves coming from the two slits are superimposed on the screen: where crest meets crest you get a crest twice as big and four times as powerful; where crest meets trough you get nothing—a dead spot. So, once again, it appears that electrons behave as classical waves. And what is so strange about that, except for the fact that we have been so seriously mistaken, and for such a long time, in thinking of electrons as particles?

But just as the shrewd detective on his way out of the room unexpectedly turns on his heels with 'Just one more question...' to challenge a suspicious alibi (the suspect still sighing with relief), so you should turn around and ask: how can the electrons be waves if they always arrive *whole* on the back screen? To see how absurd the wave proposition is, imagine you have reduced the flux of incoming electrons to the level that only one electron at a time goes through the apparatus. With great patience we can still map out the spatial distribution of the incoming electrons, and we find that it still has the maxima and the minima shown in Fig. 63. But now there is no way to interpret what we see as the interference of two waves: there is one and only one electron in the apparatus at any given time.

Very peculiar...A classical wave, no matter how faint, is always distributed in space: think of waves breaking along the seashore. Not so for electrons. Here we have a small lump of matter that can be found only at one spot or another at a given time—unquestionably a particle. The mystery is that the probability of detecting the electron seems to be controlled by a wave, but as soon as the electron is detected, the wave disappears and the particle is revealed. The failure of the two great myths of classical physics, waves and particles, could not be more dramatic. The first fails to explain the lumpiness of what is detected, while the second is shattered by the evidence that the putative particle does not go through one or the other slit, but, somehow, through both.

What is going on? Nobody knows. Every attempt to find a reasonable answer in terms of particles or waves has failed so far—or else the answer is a tightly guarded secret. As far as we know, electrons obstinately refuse to be forced into the mould of classical particles or waves. So do all the elementary particles of modern physics. They challenge the traditional conceptual system like the Moosbrugger affair challenged the Austrian system of justice, according to Musil (see Chapter 2). Faced with such a challenge one can either stick to 'pseudo-reality' (i.e. traditional particles and waves) and thus effectively forfeit reality, or embrace the fantastic, reject the conventions, recognize that nature may be working on a plan different from what we thought, and thus arrive at a higher, if somewhat incomplete understanding of reality. The second way is the way of quantum mechanics—a fantastic theory that was invented to make sense of a reality that would not be caught in the net of older ideas.

Prying into the mystery

To better understand how deeply an electron differs from a classical particle or wave, let's try to picture what would happen if we attempted to pry into the mystery of the two-slit experiment. We might put a source of light behind the screen, hoping to catch the electrons in the act of entering the apparatus. As soon as an electron comes in through one of the slits—or perhaps through both of them simultaneously—a flash of light reveals its position. This is a nice idea in theory, but it will only work if the experimental apparatus fulfils two requirements.

The first requirement I call 'tightness', meaning that every electron must be hit by a photon immediately upon entering the apparatus. It is like a tightly guarded gate in an airport: everyone must be screened, no exceptions. For this, the intensity of the light must be sufficiently high.

The second requirement I call 'sharpness'. This is needed to distinguish between an electron coming through slit 1 and one coming through slit 2: we don't want to mix them up. To achieve sharpness, the wavelength of the photon must be shorter than the distance between the slits.

If tightness fails, then many electrons will slip through unseen. If sharpness fails, then we'll see fuzzy flashes of light coming from both slits at the same time, and we'll be unable to tell which way the electrons went. When both requirements are met, the electron-watching device works as intended, but the results of the observation are deeply disappointing: the electron is found to behave like an ordinary classical particle. What happens is, light flashes are seen behind one slit or the other, but never simultaneously behind both. Then, all the electrons that strike the back screen can be neatly divided into two groups depending on whether they were observed to pass through slit 1 or slit 2. The either/or logic of particles is back in full force and tells us that the total number of electrons striking the screen is the sum of the numbers arriving separately from each slit. In brief, watched electrons do not interfere: the very same electrons, which produced wave-like interference in the 'unwatched' version of the experiment, now behave like conventional particles.

So the electrons behave in a different manner depending on whether they are being watched or not? Yes, but this is only mildly surprising. After all, we are comparing two different experiments, and there is no logical contradiction in finding different results. We all understand that the process of observing something can produce unwanted side–effects, particularly when the observed object is as small and delicate as an electron. Far sturdier objects can be affected by observation. Even a small change in the electrons' trajectories might be enough to destroy their delicate interference pattern.

Yes, all this is true and familiar, and the art of the good experimenter lies precisely in her light touch, in her ability to reduce the disturbance to such a level that it can be ignored. But, with electrons, this light touch is unattainable. No matter how good an experimentalist you are, no matter how much care you put in building your apparatus, you cannot find out which slit the electron goes through without destroying the interference pattern. The interference pattern can only be seen as long as the experimental apparatus is not sharp enough or tight enough to determine which slit the electron went through.

Up to a point, the last assertion can be understood from what we have already said about the nature of light. When a photon collides with an electron it inevitably jolts its trajectory. The only way to reduce the jolt is to make the photon 'softer', i.e. to reduce the photon's energy and momentum. This can be accomplished by increasing the wavelength of the photon, but, in so doing, we inevitably lose sharpness: light of too long a wavelength will not be able to distinguish between the two slits.

Here is how Feynman dramatizes this transition:

> Let us try the experiment with longer waves... At first, nothing seems to change. The results are the same. Then a terrible thing happens... When we make the wavelength longer than the distance between our holes, we see a big fuzzy flash when the light is scattered by the electrons. We can no longer tell which hole the electron went through!... And it is only with light of this color that we find that the jolts given to the electron are small enough... so that we begin to get some interference effect.[2]

Actually, the impossibility of catching the electron in the act of interfering with itself (or being in two places at once), is even deeper than this passage suggests. It is not just that we can't eliminate the

---

[2] R. P. Feynman, *Lectures on Physics*, Vol. I, 37–6.

disturbance. No, any process from which one could infer which slit the electron goes through would destroy the interference, regardless of whether the electron receives a jolt or not. For example, we could modify our detector so that only slit 1 is illuminated. Now when the electron goes through slit 1 we see (as before) a sharp flash behind that slit, but when it goes through slit 2 nothing happens. You might think that this detector is better than the previous one because it does not disturb the electrons which go through slit 2, so we should see at least those electrons interfere. Not at all! The reason is that the new system gives out the same amount of information as the previous one—enough to tell which way the electron goes. If you don't see any flash behind slit 1, that definitely means that the electron has gone through slit 2. And as long as you can have this certainty, you cannot observe interference. It is only by reducing the sharpness or the tightness of your apparatus that you can observe interference. For example, by reducing the intensity of the photon source you may allow a fraction of the electrons to sneak in without giving out any information about their path. Well, those electrons should show interference, and indeed they do.

So, the central question is the *amount of information* that can be gathered about the electrons with a given experimental set-up. If too much information is extracted—enough to make a sharp distinction between the two alternatives, electron going through slit 1 or through slit 2—then the interference pattern is destroyed. The difference between two light flashes coming from different locations, or between a flash and no flash, are examples of 'sharp differences'. Recording such differences is called a *measurement* in the language of the quantum theory. So a measurement need not necessarily disturb the electron: but it must create conditions that make it possible to give a sharp answer to questions such as 'did the electron go through slit 1 or through slit 2?'

According to the quantum theory, it is meaningless to say that the electron went through this or through that slit unless you have in place an apparatus which can answer the question. When you do have such an apparatus, then you don't see the interference, as logic demands. When you don't have it...well, then you can't know what happened and you shouldn't waste your time worrying about a question that can't be answered!

Still, many people are ill at ease with this dangerous logic. They feel that the electron must know in its heart of hearts what it is doing, even if we don't; even if we can't. The idea that the *non-observation* of an electron behind slit 1 affects the behaviour of that electron as it passes undisturbed through slit 2, does not go down well with these people. How can a physical effect be caused by something that did not happen? But, according to the rules of quantum mechanics, the mere *possibility* that the electron might be observed behind slit 1 makes an enormous difference—even if this observation does not actually take place. Another way of putting it is that in quantum mechanics the whole experimental set-up determines the outcome of an experiment, and this set-up, in general, cannot be separated into smaller parts. So, in our example, the behaviour of electrons that pass through slit 2 cannot be correctly described without taking into account the presence of a source of light behind slit 1. Such a theory is called 'counterfactual' because mere possibilities are as important as actual facts: Musil would have loved it.

The oracle

At their best the rules of quantum mechanics constitute a highly developed system of divination. Figure 64 shows the most famous oracle of antiquity: the oracle of Apollo in Delphi. Kings from all parts of Greece came here to receive advice about matters of life and death, such as wars, marriages, abdications. The question had to be phrased and asked in a ritual manner. The priestess, known as Pythia, would listen, ascend the altar, perform acts of worship, fall into a trance, and then, amidst screams and hysterical convulsions, release the answer in obscure,

FIG. 64 The oracle of Apollo at Delphi, Greece.

ambiguous words, whose meaning could be bent this or that way to suit the wishes of the questioner.

Up to a point, it's the same with quantum mechanics. You approach it reverently with a precise question about the outcome of an experiment. The question must be asked in the appropriate language: mathematics. Only then will the theory deign to answer you, and the answer will be a set of numbers expressing the probabilities of different outcomes. Unlike the answers of the Pythia, however, the answers of quantum mechanics are crisp and clear, and you can work them out for yourself without the help of high priests. There is no fooling around with the results, but you will never get to know what is really going on inside the physical system. Probability replaces ambiguity.

And now for the rules. Suppose you want to know the probability that the elusive quantum particle will be found at a certain point in space. The answer to your question is encoded in an abstract field known as the *wave function*. Superficially, this looks like other fields we have encountered, for example, the electric and the magnetic field.

Imagine space filled with infinitely many arrows, all perpendicular to a 'vertical' axis, but otherwise of variable length and direction. Each point in space has its own arrow attached to it, and the set of all the arrows constitutes the wave function. The wave function evolves in time like any mechanical system, meaning that the lengths and orientations of the arrows change according to strict, predetermined rules. There is no uncertainty, no probability whatsoever in this evolution. It is as determined as the best clockwork of classical physics. Probability appears only when we use our arrows to make predictions. Namely, the probability of finding the quantum particle at a given point and at a given time is simply proportional to the square of the length of the arrow at that point and at that time.

In Fig. 65, I have tried to visualize the wave function for three special states of an electron. The picture in the middle (b) represents an electron bound to a proton in the normal state (ground-state) of the hydrogen atom. All the arrows point in the same direction (the direction itself is not fixed, but it must be the same for all the arrows), and their length decreases exponentially as one moves away from the proton. 'Decrease exponentially' means that the length of the arrows is halved whenever the distance from the proton increases by a characteristic length, which in the case of the hydrogen atom is of the order of one tenth of a nanometre (nm), i.e. one tenth of a billionth of a metre. Thus it is extremely unlikely to find the electron as far as, say, 5 nm away from the proton. Such a state is therefore said to be well localized in space, but notice that even in this case the wave function extends all the way to infinity. Clearly, the electric attraction from the proton—located at the centre of the figure—succeeds in 'pulling in' the wave function, but only up to a point. Complete success would mean that the electron ends up sitting on top of the proton, just as classical physics predicts for the final state of the atom. Then the wave function would consist of a single arrow at the centre of the figure, all the other arrows having zero length. The situation is shown in the left panel of Fig. 65(a). Such a state is not impossible, but unstable. By that I mean that an electron put in such a state would fly off at a fantastic speed—attraction notwithstanding—and quickly get out of sight. The reason for this counterintuitive behaviour should become clearer after reading the next section. In short, the normal state (b) is the most compact state that can persist indefinitely. Any attempt to squeeze the electron in a smaller region of space generates a 'quantum pressure' that pushes the electron away from the proton.

The third wave function, visualized in Fig. 65(c), represents an electron that travels freely in space with definite momentum. The arrows have the same magnitude everywhere, indicating that the electron *is equally likely to be found at any position*. Now, this is very surprising: perhaps even more surprising than the failure of the electron to collapse on the proton. In classical physics, as in everyday experience, a moving particle occupies, at each time, a definite position in space. But, the wave function depicted in the figure tells quite a different story. The position of the particle is completely undetermined.

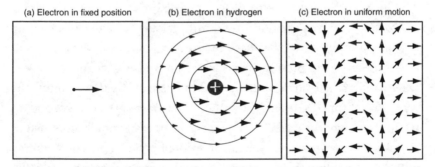

FIG. 65 Three possible states of an electron and their wave functions. In (a) the electron is localized at the centre of the square: the wave function is zero everywhere except at the point where the arrow is seen. In (b) the electron is bound to a proton (located at the centre of the square): the magnitude of the wave function decreases with increasing distance from the centre, but is never zero. Finally, in (c) we have a completely delocalized electron. The *magnitude* of the wave function is the same everywhere, but its 'phase' (i.e. the direction of the arrow) varies, giving information about the momentum of the particle.

How can we tell, from this picture, that the particle is in motion with a definite value of the momentum? A clue comes from the orientation of the arrows, which, as we see in Fig. 65, turn counterclockwise as we move to the left, but do not change orientation as we move up or down. This pattern of rotation of the arrows is the telltale sign of a quantum particle moving to the left with constant momentum. The faster the motion, the more rapidly the arrows turn. The slower the motion, the more slowly the arrows turn. A particle of zero momentum would have all the arrows pointing in the same direction. In all cases, the pattern is periodic—meaning that the arrows return to their original orientation after a fixed displacement in space. This fixed displacement is the wavelength of the electron wave.

So what happened to our ideal point particle with its definite position *and* definite momentum? We have a serious problem here. The closest approximation to that ideal is the so-called wave-packet, i.e. a wave function that qualitatively looks like the middle panel of Fig. 65. But such a state is non-stationary (i.e. non-permanent), unless there is a proton pulling on the electron. By confining the electron to a small region of space, we generate a 'quantum pressure'. In a hydrogen atom, this quantum pressure is balanced by the attraction from the proton. In the absence of the proton, that very same pressure would cause the wave packet to spread out well beyond its original size. How rapidly the wave packet spreads depends sharply on the mass of the particle and on the initial size of the packet. For an electron at rest, initially confined to a packet the size of a hydrogen atom, a doubling in size occurs within a billionth of a billionth of a second. For a stationary tennis ball initially confined to a packet the size of 1 micron, that size-doubling time would be about a hundred times the age of the universe.

The uncertainty principle

The wave function, as I have described it so far, works best when you want to know the probability of finding a particle at a certain point in space. All you have to do, then, is to calculate the square of the length of the arrow at that point. But suppose you approach the oracle with a different question like, what is the probability that this particle has a certain value of momentum? The answer to this question can still be extracted from the wave function, but with more effort. In the simplest case—a particle of definite and constant momentum—we must look at the rate of change of the orientation of the arrows (see Fig. 65(c)) and things get only more complicated for other states. On the other hand, it is possible to set up a different kind of wave function in which an arrow is attached to every possible value of the momentum, rather than to every possible position in space. The square of the length of this arrow gives directly the probability of finding the particle moving with that value of the momentum. This new wave function is called 'wave function in the *momentum representation*' and contains exactly the same information as the previous 'wave function in the coordinate representation', but responds to a different practical need: it works best when you want to know about momentum, just as the former worked best when you wanted to know about position.

Figure 66 shows again the states of Fig. 65—a particle localized at a point, a bound particle, and a freely roaming one—but now in the momentum representation. The change is striking: the two extreme cases have exchanged looks. The one that was localized in ordinary space has become very extended in momentum space, implying that large values of the momentum are possible. And the one that was extended in ordinary space has been reduced to a single arrow—a single point in momentum space. In this spectacular exchange is reflected one of the most important ideas of theoretical physics: the conjugation of position and momentum.

A state of definite momentum repeats itself periodically in space and therefore must be infinitely extended. Conversely, a state that is well-localized in space—in the sense that the ground state of the hydrogen atom is—cannot be assigned a definite value of the momentum. What about a state in which both momentum and position are simultaneously

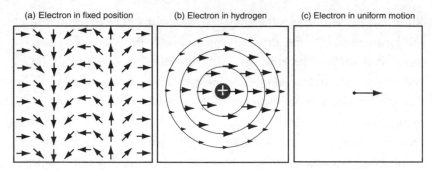

FIG. 66 The same three states as in Fig. 65, now viewed in the 'momentum representation'. Notice that the wave function of state (b) (electron bound to proton) remains more or less unchanged, but the wave functions of states (a) and (c) exchange looks. The delocalized state (c) of Fig. 65 is now described by a wave function that differs from zero only at one point in momentum space. The localized state (a) of Fig. 65 is described by a wave function of constant magnitude and varying phase.

and sharply defined? How would it look? The wave function of such a state, if it existed, would be localized in both the coordinate and the momentum representation. Then you could combine the two representations into a superior momentum *and* coordinate representation, in which the squares of the lengths of the arrows would give the probability of finding simultaneously well-defined values of the coordinate o and the momentum p. But, a look at Fig. 66 should convince you that this is impossible. The more localized the wave function is in the coordinate representation, the less localized it becomes in the momentum representation. The best you can achieve is a compromise in which neither the position nor the momentum is sharply defined, but neither is grossly uncertain. This is more or less what happens in the intermediate case of Fig. 66—the case of the electron in the hydrogen atom. If you try to further squeeze the wave function in real space, then you inevitably increase its spread in momentum space. This is the physical origin of the quantum pressure, to which I referred so casually in the previous section. Just as the pressure of a gas arises from the momentum of the molecules that strike the walls of the container, the quantum pressure arises from the extra momentum acquired by a particle that is squeezed into a smaller region of space.

Here we have the seed of a fundamental principle of quantum mechanics, *the uncertainty principle*, which rules out the possibility of simultaneously determining the position and the momentum of a particle.[3] The principle does not say that you cannot separately determine the probability of finding a certain value of q or a certain value of p, but it does ensure that you cannot calculate the probability of q *and* p having simultaneously definite values, for the simple reason that a state with simultaneously definite values of q *and* p does not exist. This is why the hydrogen atom cannot collapse below a certain minimum radius. If it did, the inevitable increase in momentum associated with the reduction in size would cause it to bounce back to a larger radius. In the end it must settle for a 'compromise' of standard size—the universal radius of the hydrogen atom, uniquely determined by the Planck constant and by the electron mass and charge. By similar arguments we can explain why in the two-slit experiment we lose the interference pattern as soon as we know which slit the electron went through. This will be done shortly.

Complex numbers

At this point I should expect to be buried under an avalanche of questions. First of all, why should we use arrows rather than plain numbers to describe probabilities? This is a deep question. The answer is that I have been using arrows as a symbol (or, if you prefer, as a representation) of something more abstract, which is the stuff that wave functions are made of: a *complex number*.

---

[3] It is a fascinating question which pairs of variables can be simultaneously determined and which cannot. The beautiful answer is that pairs of variables A and B, whose operators (see the section 'Oscillators of light' in Chapter 8, page 187) commute with each other (i.e. are such that AB = BA) are 'compatible', and can be simultaneously determined. But pairs of variables whose operators do not commute with each other, like position and momentum, are fundamentally incompatible, and cannot be simultaneously determined.

A complex number is a pair of real numbers x and y, which, taken together, form the complex entity x + iy. x and y are called, respectively, the real and the imaginary part of the complex number x + iy. The symbol 'i' is the 'imaginary unit' and has the property that its square is -1: $i^2 = -1$. Obviously there is no real number with this property, for the square of a real number is always positive.

I must confess that I laughed when a dear friend of my boyhood came up with the news that there was a number, i, whose square was -1. I thought he was making fun of me, as he often did. Lacking my southern-European levity, young Törless of Musil's homonymous novel reacted to the same revelation by falling into a deep intellectual crisis: the metaphor of the bridge consisting only of the first and the last pillar, quoted at the beginning of the book, is just a sample of his musings on the subject. Imaginary numbers seem to have impressed Mikhail Bulgakov too, judging from the book he kept on his desk during the writing of *The Master and Margarita*. The title of the book is[4] *The Imaginary in Geometry*, by the Russian mystic, philosopher, mathematician, theologian, electrical engineer, etc. Pavel Florensky, who was executed by Stalin's secret police in 1937, apparently for his refusal to reveal the place where the head of St Sergii Radonezhsky was hidden. The book claims that the imaginary geometry predicted by the theory of relativity for bodies that move faster than light is the geometry of the kingdom of God!

Be that as it may, it soon became clear that the imaginary unit was not a joke, but a deadly serious business. And the relation between complex numbers and arrows in a plane emerged in due time: every complex number can be represented by an arrow with projections x and y along the x and the y axes respectively, and, conversely, every arrow defines a complex number x + iy. Purely real numbers have y = 0 and are therefore represented by horizontal arrows. Purely imaginary numbers

---

[4] Lesley Milne, *Mikhail Bulgakov—A critical biography*, Cambridge University Press (1990).

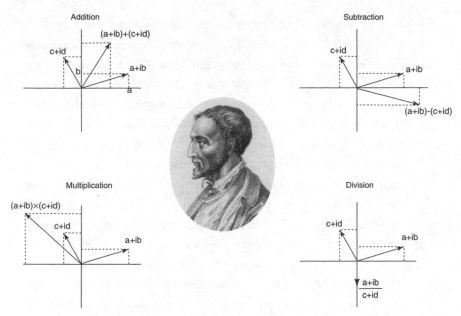

FIG 67 The four operations of elementary arithmetics for complex numbers a + ib and c + id. At the centre is Girolamo Cardano, whose formula for the solution of cubic equations first showed how complex numbers could form a bridge between real quantities.

have x = 0 and are represented by vertical arrows. All the operations of elementary algebra (sum, subtraction, multiplication, division), and many more, can be performed on complex numbers or on the corresponding arrows. For example, the sum and the difference of two complex numbers is done by combining the corresponding arrows head to tail or tail to tail, as shown in Fig. 67. So the question asked at the beginning of this section can be rephrased as follows: why is the wave function of a quantum state described by complex numbers? Why don't real numbers suffice?

The basic answer is that the two components of a complex number (or, equivalently, the length and the orientation of the representative arrow) carry information about the two conjugate variables, position and momentum—the double helix of theoretical physics. Both are necessary to generate a time evolution. For example, the wave function of Fig. 65(b), where the arrows have constant direction and variable length, describes an electron that is localized near the proton, having, on average, zero momentum. On the other hand, the wave function of Fig. 65(c), in which the arrows have constant length and variable direction, describes a particle that moves to the left with steady momentum. The length of the arrow tells us where the particle is likely to be found, the varying direction tells us how fast it is moving.

Complex numbers also help us to understand in what sense the wave function is a wave. The most important property of waves, introduced in Chapter 7, is the 'superposition principle', which says that two waves can be added together—superimposed—to produce a third wave. When a crest meets a crest you get a larger crest; when a crest meets a trough of equal size you get zero; and then there are all the intermediate possibilities. In short, a wave is described by two real numbers—an amplitude, telling us what the maximum height of the wave is, and a 'phase', telling us at what point of the cycle—crest, trough, or somewhere in between—we are. The superposition principle is also the fundamental principle of quantum mechanics. Wave functions can be added together for the simple reason that they are made of complex numbers, and complex numbers are things that have an amplitude and a phase—the length of the arrow and its direction—and can be added like waves.

Wave functions in action

Consider, for example, the superposition of two waves originating from the two slits of the two-slit experiment. This state arises when a single particle of definite momentum, described by the plane wave function of Fig. 65(c), finds its way through the screen with two slits in it. As long as the experimental apparatus does not reveal which slit the electron is going through, we must assume that the electron is in an equal superposition of the two waves emerging from the two slits. Each wave has its own set of arrows/complex numbers, and the combined amplitude at

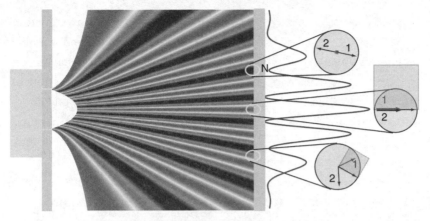

FIG. 68 The two-slit experiment explained.

each point is the vector sum of the arrows from the two waves. Look at points along the back screen in Fig. 68. At the centre of the screen the arrows from the two incoming waves point in the same direction and their sum doubles their length. The probability of striking the central point is proportional to the square of the length of the resultant arrow and is therefore four times as large as the probability of striking that point when only one slit is open. Consider next what happens at the 'dead spot' N. Here, the arrows from the two incoming waves point in opposite directions and cancel out. The probability of an electron striking the screen at point N is, therefore, zero. This is how quantum mechanics 'explains' the outcome of the two-slit experiment.

Figure 69 illustrates the workings of a more sophisticated apparatus: the *Mach–Zender interferometer*. Here we have a single particle (usually, but not necessarily a photon) striking a half-silvered mirror. A half-silvered mirror is a device that can, with equal probabilities, let the photon go through or reflect it at 90 degrees from the original direction, but you cannot tell which of the two possibilities is realized until you look ... As long as you don't look, the photon's wave function remains an equal superposition of two branches, the transmitted one, T, and the reflected one, R. The transmitted branch of the wave function goes undisturbed

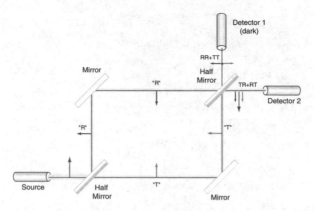

FIG. 69 The Mach–Zender one-photon interferometer. Detector 1 will *not* register as long as there is nothing in the apparatus that would allow one to determine which of the two alternative routes the photon takes.

through the glass. The reflected branch of the wave function also propagates normally, except for the fact that the reflection at the mirror imparts an extra 90° counterclockwise twist to the arrows. The two branches R and T continue to propagate along widely separated paths (Fig. 69), bounce off regular mirrors—both suffering 90° twists in the process—and rejoin at a second half-silvered mirror. At this point each branch undergoes further branching: T splits into TT and TR, depending on whether it is transmitted or reflected, and R similarly splits into RR and RT, depending on whether it is reflected or transmitted. Finally, RR and TT are superimposed in photon detector 1, while RT and TR are superimposed in photon detector 2. But the arrow of the RR branch has suffered two 90° twists—two reflections—and is therefore pointing opposite to the arrow of the TT branch: this means that the two contributions cancel, and the probability of the photon arriving at detector 1 is zero! So the outcome of this measurement is (ideally) 100 per cent certain: detector 2 will always register, in spite of the fact that the half-silvered mirrors exert an unpredictable influence on the photon.

Paradoxically, you will have this certainty only as long as you cannot tell which path, R or T, was taken by the photon en route to its final destination. Imagine that you insert an obstacle along the path of the T branch. Now you know with certainty that a photon arriving at a detector must have taken the R path, because the T path was blocked. But precisely for this reason you can no longer tell which of the two detectors will register: either can, with 50 per cent probability.

How can the presence of an obstacle along a path *not taken* influence so radically the behaviour of a photon that took a different path, and which presumably never knew of the presence of the obstacle? That's where the mystery of quantum mechanics lies. And there is nothing I can say to demystify it. Words attempt the task and come back defeated.

A classical wave represents energy distributed in space, so that two detectors placed at different points A and B may register simultaneously its arrival. But the quantum wave function represents the probability that one and only one particle manifests itself at one position and time. Two detectors placed at different points will never register simultaneously the arrival of the particle. In fact, if one registers, you can be sure that the other will not. The wave function exists only in an oracular capacity. Its purpose is to predict the probability of *future* observations. As soon as the observation is made, the original wave function collapses—its job done—and is replaced by a completely different wave function that describes the new state of affairs. If, for example, the particle is observed at point A, then the new wave function, after the observation, will consist of a single large arrow at A. So, the time evolution of a quantum mechanical object is pictured as a succession of smooth rotations interrupted by sudden, unpredictable jumps—called measurements—in which the wave function collapses on a single classical possibility.[5]

---

[5] Physicists are still grappling today with the problem of 'quantum jumps', also known as the 'measurement problem'. The idea of quantum jumps has elicited strong and diverse reactions from outstanding physicists. Some hated it, like Albert Einstein, who declared: 'God doesn't play dice', and Erwin Schrödinger, who said: 'If these jumps were to stay, I would have never started the damn thing'. To Paul Dirac, on the other hand, the quantum jumps formed 'the uncalculable part of natural phenomena' and the key to explaining the complexity of the universe.

It is not easy to attribute an objective reality to something that fills the space at one moment, only to collapse to a point at the next moment. It all seems to depend on the amount of information that is available, rather than on the particle itself. But to the young founders of quantum mechanics the existence of an objective reality was not the most important feature of a good theory. To them, the oracular virtue of the wave function was the true miracle.

The quantum state

It should be clear, at this point, that whether quantum mechanics admits an objective reality or not, we cannot hope to find that reality in the wave function itself. A wave function is, at its very best, a convenient representation of the underlying reality—one of many possible and equivalent representations, which are devised to emphasize this or that property of the system, e.g. position or momentum. So, if there is any objective reality to talk about, it must be in the abstract quantum state, which underlies all possible wave functions. This abstract state can be thought of as a vector in an infinite-dimensional space. Here you must stretch your imagination to see something like the ordinary space with infinitely many mutually perpendicular(!) axes. The vector is specified by complex components along the axes, just as ordinary vectors are specified by their real components along three perpendicular axes. The set of these components constitutes *a* wave function. For example, the infinitely many mutually perpendicular axes can be chosen to correspond to the infinitely many possible positions of a particle in space. Then the components of the state vector along these axes give the ordinary wave function—a set of complex numbers, one for each position. If, instead, the mutually perpendicular axes are chosen to correspond to the possible values of the momentum of the particle, then the components of the state vector along these axes give the momentum wave function—a set of complex numbers, one for each momentum. Choosing different sets of mutually perpendicular axes is like changing the orientation of the reference frame: this changes the wave function, but not the state itself—just as walking around an object changes the appearance but not the nature of the object.

It is not easy, in general, to visualize what I have been saying in the previous paragraph, but there is one example in which the abstract space of quantum states luckily coincides with the familiar space of vectors in a plane. I am talking of the polarization states of the photon.

Let us go back for a moment to the classical electromagnetic waves of Chapter 7. There are two basic independent types of electromagnetic waves: one in which the electric field oscillates 'up and down', while the magnetic field oscillates 'in and out' of the page, as in Fig. 49, and one in which the electric field oscillates in and out and the magnetic field up and down. These two types of wave are called *vertically polarized* (**V**) and *horizontally polarized* (**H**) respectively. Every electromagnetic wave can be constructed from a combination of these two basic types. For example, an electromagnetic wave in which the electric field vibrates at a direction halfway between the vertical and horizontal (at $45°$ with either) would be designated by **V** + **H**. An electromagnetic wave in which the electric field turns in a circle around the direction of propagation—a *circularly polarized wave*—would be denoted by **V** $\pm$ i**H** with a + or a − sign depending on whether the rotation of the field is clockwise or counterclockwise.

All I have been saying above for electromagnetic waves remains true for individual photons—the elementary quantum particles of light. A **V**-photon is, by definition, a photon with the property that it is transmitted with absolute certainty (100 per cent) by a vertically oriented 'polaroid' filter and absorbed with equal certainty by a horizontally oriented 'polaroid' filter (Fig. 70, right panel). An **H**-photon behaves in the complementary way: it is definitely transmitted by a horizontal filter and definitely absorbed by a vertical one. A (**V** + **H**)-photon, on the other hand, has a 50 per cent probability of being transmitted and a 50 per cent probability of being absorbed by either a vertical or a horizontal filter. You see that every orientation of the 'polaroid' filter

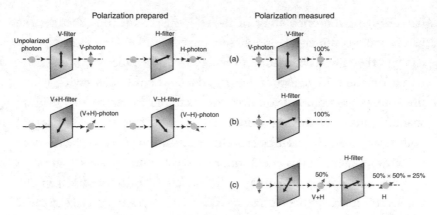

FIG. 70 Left: States of linear polarization of a photon are prepared by passing the photon through a 'polaroid' filter. The filter has an 'easy' axis, whose orientation determines the direction in which the emerging photon will be polarized. For example, vertically polarized photons (abbreviated V-photons) are generated by passing photons through a filter whose easy axis is vertical (a V-filter). Similarly, horizontally polarized photons (abbreviated H-photons) are generated by passing photons through a filter whose easy axis is horizontal (an H-filter). Right: V-photons have 100% probability of transmission through a V-filter, and 0% probability of transmission through an H-filter. If, however, a V-photon is passed through a (V + H)-filter—i.e. a filter whose easy axis forms a 45° angle with the vertical and horizontal directions—then its polarization state is reset to (V + H). In this state, the photon has 50% probability to pass through a V- or an H-filter.

defines a special quantum state of the photon—that is to say, a state that is certainly transmitted by a filter oriented along that direction and certainly absorbed by a filter oriented in the direction perpendicular to it. In this sense you may say that these quantum states are described by vectors in the (**V**,**H**) plane, with the understanding that only the direction of the vectors matters, not their length or even their actual orientation: **V** and − **V** represent the same state.

But how do we know that a photon is, say, in the **V** state? In the next section we'll see that the only way to know it for sure is to deliberately prepare it in that state. A photon of unknown origin may well be in the **V** state, but we cannot prove that it is. In the absence of additional information we should assume that it is 'unpolarized' and represent it without an arrow. It is only after the photon has been successfully transmitted through a **V**-filter, or an **H**-filter, or a (**V** + **H**)-filter that we can say with confidence: yes, it is now in a **V**-state, or in an **H**-state, or in a (**V** + **H**)-state. From that point on, we can begin to make predictions based on the knowledge of the state of the photon. Some of these predictions are quite unexpected. For example, we have already seen that a **V**-photon stands no chance of passing through an **H**-filter. But if you insert a (**V** + **H**)-filter between the **V**-filter and the **H**-filter something quite unexpected happens: the photon now stands a 25 per cent chance of getting it through the **H**-filter! The way this comes about is that the **V**-photon stands a 50 per cent chance of getting through the (**V** + **H**)-filter. If it gets through, however, its new state is a (**V** + **H**)-state, which stands a 50 per cent chance of getting through the **H**-filter. The combined probability of the two transmissions is 50% × 50% = 25%.

Granted: the quantum state does not tell us everything we might want to know. Most questions do not have a sharp answer, only a probabilistic one. The uncertainty principle introduces irritating restrictions on what can be accomplished. For example, it is futile to strive for a state which should be transmitted *with certainty* through a **V**-filter and also through a (**V** + **H**)-filter, because such a state does not exist. Subject to these limitations it seems, however, that the abstract quantum state vector is still our best bet for an 'objective quantum reality'. What a surprise is awaiting!

The no-cloning theorem

Suppose someone hands you a photon without telling you where she got it and what she did with it. Is there a way to find out what state the photon is in? The answer is: no, you can't. You can manipulate the photon to force it into some known state of your choosing, but you can't tell what its state was when it was handed to you.

It is not difficult to see why it must be so. Suppose we put the photon through a **V**-filter, and it goes through. Does that imply that the photon was in a **V**-state? Not at all. Most likely, it was in some superposition of **V** and **H** states, which by sheer chance happened to get through the **V**-filter. Only by doing a large number of measurements on the same photon could we infer with reasonable certainty the state of that photon.

Let me give you an example. If I throw a dice, wearing a blindfold on my eyes, and you, my trusted friend, tell me the outcome is 6, I have no reason to be surprised. If I throw it a second time, and you tell me the outcome is again 6, it would be quite unreasonable to suspect an anomaly in the dice. But if I throw it hundreds of times and I keep getting a 6, then I am justified in concluding that 6 is by far the most likely and perhaps the only possible outcome for that dice. And still, because of the blindfold, I don't know what is going on—perhaps the dice is heavily loaded, perhaps someone has printed a 6 on all the faces—but I can reasonably say that 'the dice is in state number 6', meaning that the only possible outcome of a throw is 6.

Why can't we proceed in the same way to reveal the state of the photon? Because already after the first measurement the quantum state has changed to something different from what it was, and there is no way to repeat the test, unless we happen to know what the initial state was, in which case the whole exercise is pointless. So the fundamental problem with quantum objects is that we have only *one chance* to guess their state— a single throw to establish whether the dice has a 6 on all its faces.

But, you say, there is a way around this. Why don't we make a copy of the photon, and then another copy, and then another, and keep sending these different copies through a **V**-filter or an **H**-filter until we gather enough information to make a meaningful statistical analysis? This is an excellent idea, but there is a catch—a very serious one: a quantum state cannot be copied; and a machine for duplicating quantum states does not exist in principle, let alone in practice. This damning state of affairs has the status of a theorem in quantum mechanics: the *no-cloning theorem*.

In Fig. 71, I have tried to explain why it is so. The essential point is that the quantum copier would have to obey the laws of quantum

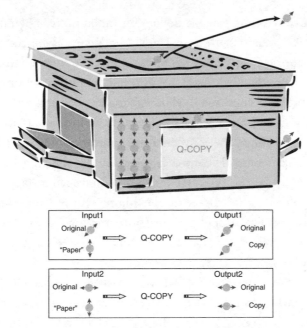

FIG. 71 The no-cloning theorem illustrated. Two different inputs (original plus blank paper) cannot be faithfully copied without changing the 'angle' between the inputs. Such a change, however, is forbidden by the basic rules that govern the quantum mechanical evolution of a state.

mechanics. According to these laws, the undisturbed time evolution of the state vector is a kind of rotation in an infinite-dimensional space—the technical name for it is 'unitary transformation'. As with every rotation, the quantum evolution preserves the angle between state vectors. So if two state vectors initially form an angle of, say, 45° with each other, then they will stay that way forever, or at least for as long as they are left undisturbed. The problem with the quantum copying machine is that it cannot do what it is meant to do without somehow changing the angle between two state vectors. You see, the quantum copying machine does not know what the state to be copied is, any more than an ordinary copying machine knows what is written in the document that you put on the glass. No matter what the input is, the machine performs a series of mechanical operations whose final result is

219

to print a copy of the original document on an initially blank sheet of paper, while leaving the original unchanged. Similarly, the task of a quantum mechanical copier is to reproduce the state of the original photon in an initially 'blank' photon (i.e. a photon prepared in a conventional reference state), while leaving the original photon undisturbed. In this case, however, the series of operations would have to boil down to a rotation of the combined state vector of the original photon *and* the blank photon. The essential point is that the angle of rotation would have to be the same for all inputs. But such a blind action is incompatible with the fact that the blank photon must be rotated through two different angles in order to produce a faithful copy of two different inputs (see Fig. 71). And this is why the quantum copier is not just hard to make: it's impossible.[6]

This is a sobering result. Just when we thought we had identified the bearer of quantum reality—the quantum state—we discover that it cannot be reproduced, and therefore cannot be reliably known. And not because we are not smart enough to figure it out: no, it's unknowable by its own unalterable decree. There is something singular and secret in a quantum state, something that only the quantum state knows and will never confess, like the cat of T. S. Eliot's poem.[7]

---

[6] Why then are ordinary copiers found in every office? Doesn't quantum mechanics apply to them? The answer is that those ordinary copiers would not be considered copiers by quantum mechanical standards. If you were to compare the original document and the copy at the microscopic level (I mean, atom by atom) you would find that they are completely different. Of course those differences do not affect the appearance of the document at the macroscopic level. This is another illustration of the ideas presented in Chapters 4 and 5: the no-cloning theorem is broken at the macroscopic level, not because quantum mechanics is violated, but because the concept of copy has a completely different meaning at that level. As a footnote to this footnote I should add that, while it is not possible to create a perfect clone of a quantum state, it is nevertheless possible to produce a high-fidelity copy of a given state. How high is high depends on what we want to do with the state: i.e. ultimately, on the adopted level of description.

[7] T. S. Eliot, 'The naming of cats', in *Old Possum's Book of Practical Cats*.

Quantum pragmatism or quantum reality?

So we are back to the question of 'objective reality' and let me digress about it. First of all, I'd like to say that the older theories based on particles, waves, vortices, celestial spheres, and so on, are not more realistic than the quantum theory: they just seem more realistic, because they are based on more familiar concepts. Those familiar concepts, when carefully analysed, turn out to be abstractions, limits, idealizations of a reality that is never completely described. True: those theories do not have dark corners like the uncertainty principle; they do not call into question the existence of objective reality; they do not force us to walk logical tightropes from which we could fall breaking our necks. But these advantages are greatly outweighed by the fact that, when applied to the microscopic world, they lead to predictions that are in sharp conflict with experiment. On the other hand, quantum theory, though perhaps philosophically irksome, has an outstanding record of successful predictions, and has never been found wrong—so far—in any application. In fact, pragmatic success was its only chance of survival, for who would have embraced such a dangerous theory, had it not been so spectacularly right in each and every prediction? From this point of view quantum mechanics could well be called 'the triumph of pragmatism', rather than 'the triumph of the abstract'.

Pragmatism has a long and distinguished tradition in the history of science. When in 1573 Copernicus' heliocentric theory of planetary motion was finally published—only days before Copernicus' death—the editor of the book was so wary of going against the established wisdom that he included a foreword, essentially disclaiming the existence of any physical reality behind the heliocentric model. Everything in the book was to be regarded as a 'mathematical hypothesis', but also as a highly practical scheme to better describe the motion of the planets. This turned out to be wise tactics, for it allowed scientific progress to take place quickly and smoothly, without getting entangled with questions of truth and authority. And perhaps there would have been no embarrassing confrontation on the heliocentric system had it not been for a strong-headed character like Galileo, who insisted that the heliocentric system must be real, the moon was not made of glass, Jupiter had its own moons, and so on. The fate of that outstanding thinker—forced to retract his theory and placed under house arrest for life—must have taught scientists a good lesson, for happily ever after they refrained, by and large, from taking a firm stance against power. So when quantum mechanics burst on the tragic stage of the twentieth century, its feet were firmly planted in a tradition of 300 years of spectacular advances, yet strangely disconnected from reality, like the works of the legendary craftsman Daedalus, who enjoyed extraordinary creative freedom, yet was locked up in a tower, so that he might not escape and spread his knowledge in the world.[8] Invisible electromagnetic waves had been predicted, then detected; invisible atoms had been hypothesized, then shown to be responsible for the manifold shapes of matter; time itself had been reshaped to comply with fundamental principles of invariance—all this without scratching the hard shell of prejudice and stolid assurance that led millions of people to march in tight ranks at the beat of a drum. And now quantum mechanics made explicit what had been implicit all along: that there was no objective reality to hold onto, that we were hopelessly lost in a game of smoke and mirrors, in a big meaningless Universe which we could perhaps control, but never understand.

But should it really be so? Or isn't quantum mechanics the beginning of a new phase in the history of human thought, one in which scientific imagination comes of age, becomes aware of its power to generate truth from fantasy, rips off the straight-jacket of narrow-minded realism and lifts itself up in the air, like Daedalus on his own wings? You must know what I think the answer is. So I will devote the next chapter to describe the positive, liberating content of quantum mechanics. No more talk of limitations and constraints! I will tell you about features of quantum reality, which expand, rather than restrict, the boundaries of the possible. I call these: tales of quantum reality.

[8] Ovid, *Metamorphoses*, Book VIII.

Tales of quantum reality

Seeing in the dark

In the preamble to his book *Escher, Gödel, Bach*, Douglas Hofstadter mentions an electric 'Black Box', whose only function was to turn itself off. The toy inspired him to a lifetime of work on self-referential systems. To most of us, however, few things are more frustrating than a device that makes itself useless as soon as we try to use it. To this class of devices belongs the ultra-sensitive bomb introduced by theoretical physicists Avshalom Elitzur and Lev Vaidman. Its detonator is so sensitive that a single photon can trigger the explosion. But precisely for this reason it is impossible to make sure that the bomb will work properly without actually exploding it—an action that leaves you with no bomb, if you are the general, and with no job, if you are the scientist who designed it. Elitzur and Vaidman, however, showed that there is a smart way to test the bomb without destroying it. This way makes use of a subtle quantum mechanical effect, which can be described as 'seeing in the dark'. The subject of this section is a pointedly non-militaristic version of their famous bomb-testing problem and its solution.

Imagine that after painstaking search you have discovered a particular form of the organic molecule tetrathiafulvalene-tetracyanoquinodimethane (abbreviated TTF-TCNQ) which cures glioblastoma—a most deadly brain tumour.[1] This particular form of the molecule occurs rarely in nature, so it is essential to devise an efficient procedure to separate the useful molecules from the common and useless ones. Direct X-ray analysis is too slow and expensive, and chemical methods denature the molecules they are supposed to select. A member of your team discovers that the useful form of the molecule absorbs photons of a certain wavelength, while the common form does not. This is exciting news! The problem is that the absorption of the photon starts a chain of events which terminates with the molecule reverting to its useless form. You are very upset. It seems almost unfair that reality should raise such high hopes, when it has no intention to fulfil them. Then you remember the weird features of quantum reality that you learned in your youth, and you come up with an elegant solution of the problem.

The molecules will be inserted, one at a time, in the lower arm of a single-photon interferometer, as shown in Fig. 72. This interferometer has been described in Chapter 9 (Fig. 69), and you will remember that, when both arms of the interferometer are free, the photon is 100 per cent sure to end in detector 2. The presence of a molecule along the lower arm of the interferometer changes the situation dramatically. If the molecule is of the common type, then nothing happens, for these common molecules do not absorb the photon, nor change its direction of travel. But when the molecule is of the useful kind, then several interesting things may happen. The first possibility is that the molecule may absorb the photon and in so doing sign its own death warrant—for the absorption

---

[1] TTF-TCNQ is indeed a molecule of great interest to physicists, but not for medical reasons.

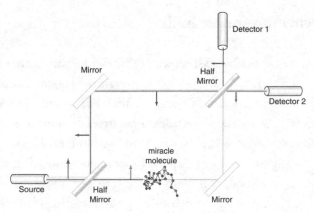

FIG. 72 A Mach–Zender interferometer enables us to 'see in the dark'.

of the photon starts a chain of events that destroys the structure. This will happen about 50 per cent of the time, that is, whenever the photon is caught travelling along the lower arm of the interferometer. In the remaining 50 per cent of cases the photon will not be absorbed, and the molecule will be left untouched, but its very presence along the path, the very fact that it *might* have absorbed the photon, is sufficient to destroy the delicate interference that prevents the photon from entering detector 1. Of course, there is still no guarantee that the photon will reach detector 1, but if it does—and the probability of this happening is another 50 per cent—then you know with absolute certainty that the tested molecule is one of the useful kind, and you can confidently put it in store. In brief, every click of detector 1 indicates that a useful molecule is present in the lower arm of the interferometer. You see why I have called this effect 'seeing in the dark'.

The irony of it all is that quantum mechanics had been presented as a theory of uncertainties, of observations that inevitably disturb the state of what is being observed. Funny that you should find a way to turn the theory against itself, to gain certain knowledge about a molecule without disturbing its state!

'A watched pot never boils'

No one has failed to observe that water boils with excruciating slowness when steadily watched, but only physicists have thought of connecting this empirical fact to Zeno's paradox, that Kafkaesque nightmare in which an arrow in flight remains nevertheless stationary at every instant (see Chapter 3). In the quantum version of Zeno's paradox, a radioactive particle, which normally would disintegrate (decay) in a fraction of a second, will simply refuse to disintegrate if you, or a mindless apparatus, keep checking its status very frequently. How frequently? The time interval between two successive checks should be much shorter than the normal decay time of the unchecked particle.[2]

To understand why this happens, notice that whenever we check the status of an unstable particle—a particle that can suddenly and unpredictably disintegrate—we are effectively asking a question, which the particle can answer with (i) 'yes I am still here', or (ii) 'sorry, I've gone'. The second outcome will be extremely unlikely if the particle is interrogated immediately after being put in place, so you can be practically sure that the first check will elicit answer (i), if it is done early enough. This argument can be applied to every successive interrogation because, according the principles of the quantum theory, every observation that elicits the 'yes' answer resets the state of the particle and erases the previous history. So after one thousand checks, all of which have been answered with a 'yes', the particle is still as good as new, and ready to go for the next one thousand checks. Needless to say, this works only as long as you keep checking, because each check

---

[2] Some experts do not believe that the scenario described in this section can be realized. On the contrary—they argue—the frequent measurements that should reset the internal clock of the particle would achieve precisely the opposite: accelerate its decay. 'A watched pot boils quicker', these experts say. The controversy is rather technical and hinges on what a real measurement does or does not do.

resets the state of the particle. As soon as you stop checking, the particle is left to its own devices, and decays within a characteristic lifetime.[3]

How to share a secret

When Captain Kidd, the legendary pirate of the Caribbean, was forced to hide his treasure chest and flee, he found no better way to remember the place where he had it buried (together with the bones of the men who buried it) than scribbling a cryptic note, which began:

53 ‡ ‡ † 305))6*;4826)4 ‡ .)4 ‡);806*;48 † 8¶60))85;l‡
(;: ‡*8 † 83(88)......[4]

Unfortunately for him, the note fell into the hands of the brilliant William Legrand, who solved the riddle in no time. The ingenious hero reasoned that, if the cipher were a simple mapping of the English language, then the most frequent symbol ('8') should correspond to the most common vowel 'e'. Next he observed that of all the words in that language, 'the' is most common: from this he deduced that ';' stands for 't' and '4' represents 'h'. And so on, step by step, until the full translation of the note appeared like invisible ink exposed to the flame:

---

[3] This curious effect has an unexpected relevance to the problem of 'seeing in the dark'. The Elitzur-Vaidman method is clever but inefficient, since it fails in 75% of the cases. The quantum Zeno effect allows us to improve the method to the point that its failure rate is only a small fraction of a per cent. The improved method is a little too complex to be described here, but the basic idea is to use the quantum Zeno effect to ensure that the photon does not 'decay', i.e. that it practically never enters the lower arm of the interferometer. 'Practically never' is different from 'never'. The mere fact that the photon might take the lower path is sufficient to make the interferometer work as a detector. The method is described in a paper by P. Kwiat and H. Weinfurter, 'Quantum seeing in the dark', *Scientific American* **275**, November 1996, p. 72.

[4] Edgar Allan Poe, *The Gold-Bug*.

A good glass in the bishop's hostel in the devil's seat...

Captain Kidd was not well versed in cryptography. Had he been better educated, he would have known that any code that closely reflects human language must yield to statistical analysis. To avoid this risk, the codes used by banks, governments, and criminal organizations are based on mathematical transformations that destroy any resemblance to human language. These codes owe their robustness to the enormous difficulty of solving certain mathematical problems, such as the factorization of a large number into prime factors (more about this later). While the encoding procedure is relatively simple, the inverse operation is so difficult that the most powerful computer could not complete it in a million years.

The difficulty of solving a problem, however, is a dangerously weak foundation for a method that purports to give absolute guarantee of secrecy. It may well be that an efficient solution to the factorization problem exists, which has escaped everybody's notice so far, but may not escape the superior intellect of a William Legrand. He could have a lot of fun and accumulate riches far greater than a pirate's treasure chest before anybody realized what had happened and started looking for a remedy. A truly secure cryptography should be based on an *absolute impossibility*, not on a practical and perhaps temporary limitation of our art.

Quantum reality offers the possibility of realizing such an absolute impossibility. We have seen that a quantum state cannot be reliably identified, except in those trivial cases in which we have complete knowledge of the process by which the state was created. In all other cases, the identity of the state cannot be reliably uncovered. In this unfathomable depth we will now bury our secret, and its secrecy will be guaranteed by the laws of quantum mechanics itself.

We begin by translating the English text of the message into a series of **o**'s and **1**'s according to some conventional code. This can be an easy-to-break code like the one used by Captain Kidd, except that the characters are short strings of **o**'s and **1**'s. Encryption is done by adding to the **o**'s and the **1**'s of our message the *secret key*—a random list of **o**'s and **1**'s, which the encrypter chooses according to her will. Thus, the encrypter adds to the first digit of the message the first digit of the key, to the second digit of the message the second digit of the key and so on. (obviously, the key must be at least as long as the message). The addition itself is not the usual one, but a special kind designed to give **0** or **1** according to the following rule: you get **0** if you add **0** and **0** or **1** and **1**; and you get **1** if you add a **0** and a **1** in any order.[5] The encrypted message is absolutely impervious to statistical analysis because each letter of the original text is converted into a random, unpredictable string of **0**'s and **1**'s. However, the message can be decrypted very easily if you know the key: all you have to do is 'subtract' the key from the encrypted message.

The communication begins with the transmission of the secret key from the sender (traditionally called Alice), who uses it to encrypt the message, to the recipient (traditionally called Bob), who will use it to decrypt the message. However, there is clearly a major problem here: what is the point of encrypting the message if the key itself is not encrypted, but freely released in the air, for anyone to listen? We need a secure way to transmit the key—or, at the very least, we must ensure that a would-be eavesdropper (traditionally called Eve) cannot copy the key without us noticing. This is what in North-American parlance is called a 'Catch-22 situation':[6] we need a secure communication channel to set up a secure communication channel. What shall we do?

Having lost faith in classical methods, we turn to quantum mechanics. As a first shot, we represent the **0**'s by vertically polarized photons (**V**-photons) and the **1**'s by horizontally polarized photons (**H**-photons).

---

[5] And still some people say: it is a fact that one plus one is two!

[6] Curiously similar to Gödel's **G**-proposition (see Chapter 2), a 'Catch-22 situation' arises when a procedure cannot be carried out because the requirements for its execution are self-contradictory. In Joseph Heller's novel *Catch-22* the situation arises when an Air Force pilot applies for psychiatric evaluation. The pilot wants to be declared insane so that he will not have to fight, but the very fact that he has applied for a psychiatric evaluation proves that he is sane.

Our secret key now consists of a series of photons, randomly polarized in the **V**- or in the **H**-direction. Unfortunately, this does not solve our problem, for the eavesdropper can steal the photons, identify them as **V** or **H**—an easy task once you know they can only be of the **V** or of the **H** kind—and finally replace them with identical photons—all without leaving a trace. To prevent this, we should hide the fact that our photons can only be of the **V** or of the **H** type. Lacking this crucial information, the eavesdropper will be left in the dark, but so will the intended recipient of your message. How do we get around this?

The problem is solved by representing each digit of the secret key in one of *two* possible ways. A **o** can be represented either by a **V**-photon (⇑) or by a D_1-photon, where D_1 is a shorthand for the diagonal polarization state **V** + **H** (⤢) depicted in Fig. 70. Similarly, a **I** can be represented either by an **H**-photon (⇒) or by a D_2-photon, where D_2 is a shorthand for the diagonal polarization state **V** − **H** (⬉) (also shown in Fig. 70). So we have two possible representations of **o** and **I**, which we call the (V,H) basis or the (D_1, D_2) basis. The idea is to switch randomly between these two representations, in such a way that the potential eavesdropper (Eve) will never know for sure whether the photon that is flying from Alice to Bob is to be interpreted in the (V,H) basis or in the (D_1, D_2) basis. Not knowing in which of the four possible states ⇑, ⇒, ⤢, ⬉ the photon is prepared, she cannot reliably read the information.

But, you say, doesn't Bob run into the same difficulty? Yes, of course, but he has a nice way around it, provided that Alice is willing to cooperate. The procedure is described below and in Fig. 73, and it is known as the *BB84 protocol*. So, here is what Bob and Alice do.

First, Bob must acknowledge receipt of each photon: this step ensures that the photons have not fallen into Eve's hands—each and every one must be accounted for. Second, he reads each incoming photon in the (V,H) or in the (D_1, D_2) basis according to the random toss of a coin (for example he may use the (V,H) basis if he gets heads, and the (D_1, D_2) basis if he gets tails). Of course, his guess will be wrong about 50 per cent of the time, yet he keeps reading and writing down a tentative list of digits. Third,

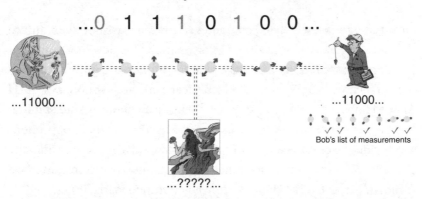

...?????...

FIG. 73 The BB84 protocol enables Alice and Bob to share a secret list of numbers. In Step 1 Alice sends Bob a sequence of 0s and 1s coded sometimes as V or H photons (1=V,0=H) and sometimes as D_1 or D_2 photons (1=D_1,0=D_2). Which of the two coding schemes is used for a given photon is decided by the toss of a coin. Bob acknowledges receipt of each photon, and interprets it as he wishes (i.e. randomly) either on the (V,H) basis or on the (D_1,D_2) basis. In Step 2 Alice reveals the sequence of encoding schemes (not the actual numbers!) She and Bob keep only the numbers written on the photons which Bob has interpreted correctly (the checked photons in the figure). Finally, to make sure that the numbers have not been altered in flight, Alice and Bob check the consistency of a randomly selected subset of numbers from the key.

Alice reveals which basis she used to write the information on each of the photons that she sent and Bob received. For example, she might announce: for photon No. 1 I used the (V,H) basis, for photon No. 2 I used the (D_1,D_2) basis, and so on. There is no need for secrecy in this step. Even if Eve listens, the information will be useless to her: those photons are gone forever, processed and destroyed by Bob's reading apparatus. Fourth, Bob checks his records and identifies the cases in which he correctly guessed the basis that Alice had used (this will be about 50 per cent of the cases). He sends this information back to Alice, publicly, and again in a way that is useless to Eve. For example, he might say: with photon No.1 I guessed right, with photon No. 2 I guessed wrong, and so on. You see that no useful information is released to Eve in this step, since she doesn't know what was written on photons No. 1 and No. 2. However, Alice and Bob now share a secret: they both know the digits that Bob read from the photons which he happened to interpret on the correct basis. This set of numbers, which they and only they know, will be their secret key—the basis of their future communication.

The final precaution is to make sure that the two copies of the key now in their possession are indeed identical, i.e. that the photons sent by Alice to Bob have not been altered in flight. To make sure, Alice and Bob compare some of the numbers from the key. For example, Bob might say: my digit No. 127 is a 1 and my digit No. 262 is a 0. Alice checks the elements No. 127 and No. 262 in her copy. Of course, she must also find 1 and 0, respectively. This test can be repeated many times, and, if agreement is found in every case, then it can be safely concluded that the two copies are identical. True, the tested digits No. 127 and No. 262 are no longer secret, but who cares? They can be discarded, while the remaining digits can be safely used as the key.

If you are like me, you will have to read the above paragraphs a few times to fully understand and appreciate the ingenuity of the method. You might ask: where did we make use of quantum reality? Why could we not implement this method in classical reality? The answer is that in classical reality Eve could make a copy of the streaming photons, then listen carefully to the discussions between Alice and Bob and go through the same moves as Bob and forge her own copy of the key. But quantum states cannot be copied. The core information that is buried in the quantum state of the photon cannot be copied and this is what protects the secrecy of the key.

Two halves of the same body

The identical twins Yin and Yang had been born united at the hip. For many years they had shared the most intimate experiences of life until one day the scalpel of a surgeon made them into two separate people. Yet, because of their long association, they were never totally separated. Like two halves of the same body they kept feeling each other at a distance and were even rumoured to be able to read each other's mind. Was this reality, or merely powerful suggestion?

Here is a cold-blooded experiment to find out. The twins sit in separate rooms, with no possibility of conventional communication between them. One of the twins is shown a card picked at random from a deck. Is the card red or black? The twin must concentrate hard on the answer to this question, which we call X. The answer has the 'value' $X = 1$ if the card is red and $X = -1$ if it is black. In the other room the twin brother is asked to guess the colour of the card that his brother has just seen. His answer, X', will again be 1 for a red card and -1 for a black one. But he has no way to physically see what his brother has just seen: he can only rely on luck, or on the ability to read his brother's mind. This experiment is repeated many times to eliminate luck. For each card we make a note of the product XX', which is $+1$ when X and X' are agreement, -1 when they are in disagreement. In the absence of a telepathic connection the *average* value of XX' (denoted by $<XX'>$) should be zero: that is to say, the number of cases in which the two answers are in agreement is roughly equal to the number of cases in which they are in disagreement. If our experiment gives a value of $<XX'>$ that is significantly different from zero, it may be reasonable to assume that some kind of extra-sensory communication has occurred. If we found $XX' = 1$ (answers in agreement) every time, that would be compelling evidence of telepathic connection.

As far as I know, the outcome of this experiment in ordinary reality is $<XX'> = 0$ (no telepathy). But, in quantum reality there is a phenomenon that eerily resembles telepathy, even though it ultimately misses some essential features of the genuine item. Einstein denounced this effect as 'spooky action at a distance', and disliked it intensely. Today, the effect is seen as a positive manifestation of a subtle condition known as *quantum entanglement*.

The quantum analogue of the 'test for telepathy' begins with the production of a pair of entangled photons—the Siamese twins of our story. Excited atoms emit light when they jump from an excited state to a lower energy state. Normally, the jump produces a single photon, but, in some cases, it is possible that *two* photons are emitted in a single

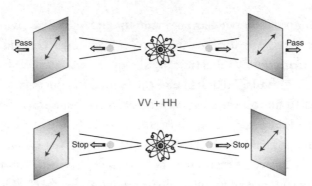

VV + HH

FIG. 74 Two entangled photons give strictly correlated answers to questions asked independently in faraway places. In this example, the two photons will either both pass or both fail to pass through identical polarization filters.

jump. Because of their common origin, these two photons will never be truly independent, even as they become widely separated in space. Their linear polarizations will always turn out to be identical. If one photon is found to be in the **V**-state, then the other is guaranteed to be in the **V**-state; if one is found to be in the **H**-state, then the other will necessarily be in the **H**-state; and similarly for any other orientation one might choose to test, such as D_1 or D_2. This correlation is far deeper than it appears at first sight, and easy to misunderstand, so let's try to be very clear about it.

You must notice that I am not saying that either photon is in a definite state of polarization. I am not saying, for example, that the two photons are, each one of them, in a **V**-state. If that were the case, then they would always pass through polarization filters oriented in the **V**-direction (**V**-filters), and would always be absorbed by **H**-filters. But this is not what happens. In reality, each photon has equal probabilities of passing or being absorbed by a **V**-filter or by an **H**-filter. The essential point is that neither photon is, individually, in a definite state of polarization. The state of the two photons, considered as a single body, is neither **VV** (↑↑) nor **HH** (→→) (the first letter refers to the first photon, the second letter to the second photon), but a *superposition* of the two: **VV** + **HH** (↑↑ + →→). The amazing thing about this superposition is that it can be written in an infinite number of equivalent ways, one for each pair of mutually perpendicular orientations: for example **VV + HH** can be also written as $\mathbf{D_2 D_2 + D_1 D_1}$ ($\nwarrow \nwarrow + \nearrow \nearrow$)—a fact that can be easily verified using the expressions for $\mathbf{D_1}$ and $\mathbf{D_2}$ and a little algebra. States of this kind are called *entangled* because they cannot be represented as the product of two independent parts. Entangled from birth, the two parts have the ability to give tightly correlated answers to random questions, even if those questions are asked independently in faraway places, with no possibility of communication between them. So if one photon is transmitted by a **V**-filter in one room, then the other photon will also be transmitted by a **V**-filter in the other room, and if one is absorbed, then the other will also be absorbed (Fig. 74). But, we cannot tell, a priori, which of the two scenarios will materialize.

Let us see how a 'telepathy test' for entangled photons might go. The two photons may travel several thousands miles in opposite directions, until they reach two labs at the opposite ends of a continent (shall we call them Guantánamo Bay and Alcatraz?) There they are trapped in special photon traps (for example, between pairs of parallel mirrors), where they can be indefinitely detained and interrogated. Now the test administrators of the two labs, always in contact via email, agree on the 'question' that will be asked. The question is whether the photon will or will not pass through a polarization filter: they agree on the direction of the filter.

Following this choice, the first interrogator inserts the filter in the path of his photon. The 'answer' is taken to be $X = 1$ if the photon goes through the filter, $X = -1$ if it is absorbed. Immediately afterwards (in fact, so soon that no information can be exchanged between the two labs) the second administrator performs the same operation on his photon. Again, the answer is $X' = 1$ if the photon goes through, $X' = -1$ if it is absorbed. The test is repeated many times, on different pairs of photons, and only when all the questioning is finished do the two administrators begin to compare the answers. Amazingly, they find

$X = X'$ (and therefore $XX' = 1$) in all cases—in perfect agreement with the predictions of quantum mechanics.

I say 'amazingly' because the photons had no way of knowing in advance which question would be asked and what their answer would be. Nor was there any communication between the two labs after a question was asked and answered in the first lab. Nevertheless the answers in the two labs always turn out to be in perfect agreement. The moment one interrogator gets a definite answer from his photon, he knows that the other photon must have jumped to a state of definite polarization, ensuring that his colleague will get the same answer in the other lab. Hard as it is to believe, it appears that the measurement carried out in one lab somehow alters the state of the photon in the other lab, even though no physical interaction is possible between the two. The two photons have somehow responded as a single body, even though they were miles apart. This is what Einstein called 'spooky action at a distance'.

I can already see many in my imaginary audience waving their hand to voice an objection. In a real telepathy test the question is asked about an objective reality, such as the colour of a card, on which the subjects have no control: there is no way the subjects can have rigged up an answer to such a question. But in the 'photon telepathy test' we ask the photon to disclose something about itself, something that we don't know but that the photon itself may have known for a long time and be perfectly prepared to answer. It is possible that each photon hides in its heart of hearts some private information—a *hidden variable*—which, if known to us, would unambiguously determine the outcome of all the measurements. If this were the case, then the probabilistic character of quantum mechanics would reflect only our ignorance of reality, not reality itself. Perhaps different photons carry different hidden variables and therefore respond differently to the same question. Being ignorant of the hidden variables, we would be tempted to interpret the outcome of the experiments as probabilistic behaviour, when in fact everything is completely determined. Then, when two photons have a common origin, as in our telepathy experiment, their hidden variables would be correlated from birth, and the 'amazing' entanglement would show its true colours: a trivial consequence of pre-existing correlations between hidden variables—a rigged-up game indeed.

This was, in essence, Einstein's point of view. He thought that quantum mechanics was not wrong but incomplete, and that the uncertainty of its predictions would go away if one could closely follow the evolution of the hidden variables. Even more importantly, the existence of hidden variables would help him to fence out that semi-mystical idea of action at a distance, which was strangely coming back in the wake of the most revolutionary developments in physics since Newton. For no matter what the hidden variables were or did, they had to determine the fate of the particle *locally*. To Einstein, this was a non-negotiable point—one of those strong prejudices which inspired his work on relativity (Chapter 6). Locality meant that the photon must be self-contained, that it must carry within itself all the information needed to determine its own fate—absorption or transmission—and that this fate should not be influenced by what was happening to another photon far away. So while the prestige of quantum mechanics was growing by the day, the father of modern physics—the man who had introduced the photon and revolutionized the concept of time—clung to the old-fashioned idea of objective reality (the hidden variables) and flatly rejected the possibility that two separate parts of a body could exert a direct influence on each other without the mediation of something going in between.

Bell's inequality

Who knows how long the issue would have remained a subject of idle dispute if a young physicist, John Bell (the same Bell of the relativistic 'paradox' illustrated in Fig. 37) had not come up with a brilliant idea to test whether quantum entanglement could or could not be explained by pre-existing correlations between local hidden variables. His idea was simple and yet very subtle.

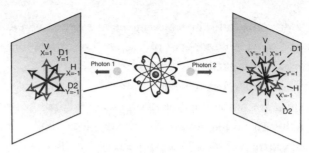

FIG. 75 In an experiment designed to test Einstein's ideas about the possible existence of hidden variables, one photon of an entangled pair is put *either* through a V-filter (vertical polarization passes) *or* through a D1-filter (polarization at 45° with the horizontal passes). At the same time, the other photon of the pair is put through *either* of two filters oriented halfway between V and D1 and between D1 and H, respectively. Thus, there are four possible measurements that can be performed on the photon pair. Like a secret set of directions, hidden variables would prescribe, from the beginning, the outcome (pass-pass, pass-no pass, no pass-pass, or no pass-no pass) for each of the four measurements. John Bell showed that no such scheme of directions can explain the strong correlation that exists between the partners of a quantum-entangled pair.

Let us go back to our photon telepathy experiment, but now let us imagine that each test administrator can ask only one of two questions. The questions allowed to the first administrator are (1) Photon, will you pass through this polarizer filter oriented along the vertical axis (a **V**-filter)? or (2) Photon, will you pass through this filter oriented at an angle of 45° to the vertical (a **D₁**-filter)? The answer to question 1 is called X and can take either the value 1 (yes) or −1 (no). Similarly, the answer to question 2 is called Y and can take the values 1 (yes) or −1 (no).

The questions allowed to the second administrator are slightly different. Question 1 asks if the photon will pass through a filter oriented halfway between **D₁** and the vertical. And question 2 asks if it will pass through a polarizer oriented halfway between **D₁** and the horizontal (these orientations are shown in Fig. 75). The answers to these questions are called X′ and Y′, respectively, and take the values 1 for yes and −1 for no. The experiment is repeated many times, with the two administrators asking their questions randomly and independently: a long list of answers X, Y, X′, Y′ is compiled, and the averages $\langle XX'\rangle$, $\langle XY'\rangle$, $\langle YX'\rangle$, and $<YY'>$ are computed. Finally, the quantity $S = <XX'> + <YX'> - <XY'> + <YY'>$ is calculated. This, of course, is nothing but the average of the quantity $XX' + YX' - XY' + YY'$.

What is the point of the exercise? Suppose for a moment that the game is 'rigged up', so that each photon knows in advance what its answers will be. This is like saying that the variables X, X', Y, Y' all have pre-assigned values 1 or -1. Then the quantity S defined above can only be 2 or -2 (check to believe!) And this implies that its average value, S, cannot be larger than 2 or smaller (i.e. more negative) than -2. This simple observation is an example of a *Bell inequality*. Its importance lies in the fact that it sets limits on the amount of correlation that can be 'rigged up' through local hidden variables. No scheme of local hidden variables can produce a value of S larger than 2. No scheme can produce a value of S smaller than -2. Any correlation that does not fall within the range -2 to 2 is beyond the power of classical reality to produce. The big surprise is that quantum mechanics violates the Bell inequality by a significant margin. For example, if the experiment described above is carried out with strictly entangled photons, quantum theory predicts $S = 2\sqrt{2}$, which is about 2.8, in glaring violation of Bell's inequality.

The experiment has been done, and Nature has shamelessly sided with the 'spooky' quantum mechanics against the hidden variables and their illustrious proponents. In other words, values of S distinctly larger than 2 have been measured. This implies that, as far as we can tell, there are no hidden variables (more accurately, no *local* hidden variables). The answers X, Y, etc. are not predetermined but, in some mysterious way, acquire a definite value—yes or no—only when the question is asked. But the biggest surprise of all is that the question asked of one part of the system will determine the answer of another part, even if those two parts are so far away that they cannot possibly interact during the timeframe of the experiment.

To some people the existence of remote correlations is a disturbing feature of the quantum theory. To others, the idea that we are part of a deeply interconnected universe is a source of endless fascination. But whether you like it or dislike it, the fact is that quantum entanglement has already opened up real possibilities for computation and communication. One of the most charming feats in communication has been the experimental demonstration of *quantum teleportation*—a technique for the disembodied transmission of a quantum state, which I now describe.

Teleportation

We have seen that a quantum state that is presented without a description of the process by which it was prepared cannot be copied nor reliably identified. So it seems that the only way to move such a state from a place to another is to physically *take it there*. You have to delicately lift the physical system in which the state resides, put it in a box labelled 'extremely fragile', and hope that it will arrive at destination without damage. There seems to be no alternative to *transportation*. Surely you cannot fax or email something that you cannot copy, much less describe in words. But appearance is deceptive. It turns out that there is an ingenious method by which we can transmit the unknown quantum state without its physical support. This method is called *teleportation* and makes creative use of entangled pairs of particles.

Let us say that Alice wants to send Bob an unknown quantum state, initially written on a photon P. The premise of teleportation is that Alice and Bob already share an entangled pair of photons, called A and B respectively, and can talk to each other by telephone. This is reasonable: after all, no form of communication would be possible between people who did not already share something, e.g. a language. Alice's idea is to let photon P interact with her own member of the entangled pair, A. As a result of this interaction the original state of P will be irreparably destroyed, but the state will be imprinted, via quantum action at a distance, on Bob's photon, B. Actually, what is imprinted on B is not exactly the original state of P but, in general, a rotated version of it (remember: a quantum state is like a vector in an abstract multidimensional space). By looking at the combined state of P and A after the interaction, Alice can determine what that rotation is and inform Bob over the phone. Finally, in the peace of his lab, Bob will undo the rotation and recover the original state of P. Teleportation is complete.

You will notice that, in spite of the instantaneous nature of the quantum action at a distance, the teleportation process is not at all instantaneous: some crucial information must reach Bob by phone, and this takes time. This is consistent with the theory of relativity, which sets an upper limit to the speed of communication. It is also consistent with the no-cloning theorem of Chapter 9, because, at the end of the process, there is still only one copy—'printed' on B, and still unknown—of the state that was initially 'printed' on P.

A schematic picture of a teleportation line is shown in Fig. 76. It is really a rudimentary quantum computer (see next section). The interaction between P and A forces an entanglement between these two photons. There are actually four types of entanglement, which are distinguished in Fig. 76 by different letters.[7] Because A was initially entangled with B, the new relationship between P and A forces an entanglement between P and B (anyone who has been involved in a *ménage à trois* will understand this). The form of this entanglement is entirely controlled by the type of entanglement that exists between P and A—the colour that lights up on Alice's dashboard. As soon as she sees it she runs to the phone and cries: it's red! Then Bob knows what type of entanglement exists between his photon and the original P, and by pushing the appropriate lever on his state-rotating machine recovers the original state.

---

[7] These four types of entanglement are known as *Bell states*: (i) **VV** + **HH** (ii) **VV** − **HH** (iii) **VH** + **HV** (iv) **VH** − **HV**, and differ in the specific form of the correlation between the photons. The difference between (i) and (ii) or between (iii) and (iv) is particularly subtle, and can be revealed only by filters that are neither vertical nor horizontal. Bell states can be created by measurements, known as coincidence measurements, which act on two photons at the same time.

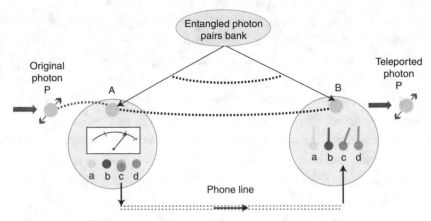

FIG. 76 Quantum teleportation is a process for destroying and reconstructing in a remote destination an object that we cannot fully describe or copy. In this cartoon, the original photon P, whose state is unknown, is forcibly entangled with partner A of a previously entangled pair of photons (A,B). Although the original state of P is destroyed in the process, the information about that state is imprinted on photon B. The owner of B can then reconstruct the original state, provided that she knows the nature of the entanglement between the original photon and A. Four possible types of entanglement are schematically indicated in the figure by letters a, b, c, d. In this example, the information that 'c' type of entanglement has occurred is communicated to the owner of B via a conventional phone line.

Quantum computation

I said that quantum mechanics enables us to create secret codes that cannot be broken, but I forgot to say that physicists are generally far more interested in divulging secrets than in protecting them. To our quintessential theoretical physicist, Richard Feynman, Nature itself was the safe to be cracked. As a young man working in Los Alamos on the ultra-secret atomic bomb project, he entertained himself cracking real safes in his free time—much to the dismay of his superiors, I presume.

The quantum computer—an elegant theoretical idea[8] to which Feynman gave seminal contributions—provides (among other things) new

---

[8] Although the theory of quantum computation is quite mature, no one has yet succeeded in building a quantum computer even remotely comparable in power to the computers we are familiar with.

possibilities for breaking into secrets. You will remember how modern schemes of cryptography are based on the enormous difficulty of factoring even relatively small numbers—say a mere few hundred digits long—into prime factors.[9] Well, I already pointed out that this method is risky, for someone might come up with a method for doing efficient factorizations. Now I can add that such a method already exists, and could be put to immediate use if one had at one's disposal a futuristic device, which is called a 'quantum computer'.

But, what is a quantum computer? Computers come in all sizes, shapes, and designs, but deep inside are all incarnations of the same abstract idea: the universal computing machine. This ideal machine is meant to work on an input that consists of two items: (1) a list of 1's and 0's, known as *input data* and (2) a list of instructions, also expressed as a list of 1's and 0's, for manipulating those data. Item (2) is the *program*. The word 'manipulating' describes any operation that one might perform on the data, such as copying, erasing, changing 0's to 1's and 1's to 0's, as well as writing new data. The machine is supposed to have unlimited resources for doing all this, and in modern computers this typically means huge expanses of microscopic switches, which can be on or off and mean 1 when they are on and 0 when they are off. Never, during the execution of a program, must the machine be left in doubt as to what to do next. Such a doubt, if it ever arose, would be considered a fatal programming error. The instructions specified in the program, together with the input data and the present state of the machine, must unambiguously determine the next step to be taken.

The above description of a computer is as classical as it can be: each bit of input, each switch, can only be in one of two states: on/off, 1/0). What

---

[9] Primes are numbers that can be divided exactly (i.e. without remainder) only by themselves and, of course, by 1. For example 3 and 7 are primes, but 21 is not, because it can be divided exactly by 3 and by 7. However $21 = 3 \times 7$ is a product of two primes, and there is no other way to express it as a product of primes. This is a general fact: every number that is not a prime can be expressed as a product of primes in one and only one way. Finding this way is called 'factoring a number into prime factors'.

could make computers different in the quantum world? The answer is that the basic quantum switch, abstractly referred to as a quantum bit or *qubit*, does not have to be either on or off, but can exist in any superposition of these two states. If, for example, our qubit is a polarized photon, we know that the two basic polarization states, **V** and **H**, are not the only two possibilities: **V** + **H** and **V** − **H**, or, for that matter, any combination of **V** and **H**, are equally legitimate possibilities. Even more important is the fact that different qubits can be entangled to create pairs, trios, quartets ... groups of qubits which respond to external action as a single entity. Herein lies the greatest potential of the quantum computer, for the twin features of superposition and entanglement enable it to operate simultaneously on all the elements of an immense batch of data—a feature known as *quantum parallelism*.

To understand the import of that last sentence, let us imagine that the qubits of our computer are photons, and let us agree that polarization **V** stands for logical state **1**, while polarization **H** represents logical **0**. Now if you have N photons, then you can represent any integer number between 0 and $2^N - 1$ (extremes included) by a particular sequence of **V**- and **H**-photons. For example the number 0 (i.e. zero—not to be confused with logical state **0**) is represented by a sequence of N **H**-photons, while the number $2^N - 1$ is represented by a sequence of N **V**-photons. In this manner any integer number between 0 and $2^N - 1$ is represented by just one sequence of **V**- and **H**-photons, according to the binary representation of the integers.[10]

---

[10] The binary representation of numbers is very similar to the decimal one, except that there are only two digits, 0 and 1. In the decimal system the value of a digit depends on its position: the last digit on the right counts for itself, the one immediately on its left must be multiplied by 10, the next one on the left must be multiplied by $10^2 = 100$ and so on: the number is the sum of the contributions from all the digits. Similarly, in the binary system the last digit on the right counts for itself, the one immediately on its left must be multiplied by 2, the next one by $2^2 = 4$ and so on. Thus, for example, the number 97 is represented by 1100001, meaning that $97 = 1 \times 2^6 + 1 \times 2^5 + 0 \times 2^4 + 0 \times 2^3 + 0 \times 2^2 + 0 \times 2 + 1$.

The computer will perform a series of operations on the input photons, creating an output that is a certain function of the input, for example the square of the input.

Up to this point everything is still classical. The photon, however, can be in any superposition of **V** and **H**, 'logical **1**' and 'logical **0**'. This opens new possibilities. Starting from a row of N **H**-photons, let us rotate the polarization of each photon by an angle of $45°$. This means that each photon is put into the state **V** + **H**. Accordingly, the input **HHHH**... (corresponding to the number 0) is transformed into (**V** + **H**)(**V** + **H**)(**V** + **H**)... Expanding the product we see that we have a superposition of all the possible combinations of **V** and **H** in any possible order. In other words, this input is a superposition of all the 2^N inputs representing integers between 0 and $2^N - 1$. But this huge superposition of inputs (2^N entries) has been achieved with a very small number of operations (N rotations, one per qubit). Any computation that could be applied to a single number (for example computing the square) can now be applied simultaneously to all the numbers between 0 and $2^N - 1$: the output will be a superposition of $2^N - 1$ computations. Herein lies the tantalizing power of the quantum computer.

The word 'tantalizing' is particularly apt. As a punishment for his rowdiness Tantalus was condemned to stand in a pool of water beneath a fruit tree with low branches. Whenever he reached for the fruit, the branches raised. Whenever he bent down to get a drink, the water receded. Similarly, it is impossible to extract but a tiny drop from the huge amount of information contained in the output of a quantum computation. The curse is that reading the output state constitutes a 'measurement', which causes the wave function of that state to collapse on one and only one of the huge number of classical possibilities. Then we get to know the outcome of *one* of the many computations that were performed in parallel, but we completely lose track of all others. An immense treasure of potential knowledge is volatilized.

In some fortunate cases, it is possible to determine a global feature of the output state, and this feature may happen to contain the answer to an important question. The most striking example known so far is the procedure devised by Peter Shor to factor a large number into primes. This is also a spectacular example of the power of global features—the spectral components of reality, as Musil called them—as opposed to a punctual knowledge of the facts. I cannot present in detail how a quantum computer would allow us to factor large composite numbers in a relatively short time, but I can try to convey the basic idea, which is fascinating.

First of all, we must realize that the factorization problem bears an uncanny relation to another problem, which seems at first devoid of any practical use, just an idle game for rainy Sunday afternoons. Take two numbers r and N with no common prime factors, and r smaller than N. For example, r = 7 and N = 9 have these requisites. Then r divided by N is 7 divided by 9, which is zero with a remainder of 7. Next consider r^2 divided by N, that is 49 divided by 9, which is equal to 5 with a remainder of 4. Let us continue with higher powers of r. The next item is r^3 divided by N, that is 343 divided by 9, which equals 38 with a remainder of 1. But if you take one more step and calculate r^4 divided by N, i.e. 2401 divided by 9, you get 266 with a remainder of 7 again. From this point on, the remainders of powers of r repeat themselves in a cycle: we get 7, 4, 1, 7, 4, 1, 7, 4, 1 ... over and over again. The *length* of this cycle (three items in this case) is called 'the order of the number 7 with respect to 9'. Every number r has some definite order with respect to any other number N. The amazing thing is that this order (the length of the cycle) is intimately related to the prime factors of N. In our example, the length of the cycle is 3, and 3 is also a prime factor of 9!

It turns out that the problem of factoring large numbers can be solved in a relatively simple way, given an efficient method for finding the order of r with respect to N. Unfortunately no such method is known—not if all we have at our disposal is classical computers. But with a quantum computer it would be a different story! The nice thing about the order-finding problem is that it does not matter what the numbers in the cycle (7, 4, 1 in the present example) are: only the length of the cycle matters. And this global property of the set of remainders can be determined without knowing what the remainders actually are.

In order to calculate the order of r with respect to N a quantum computer would first generate a quantum state that hides in its guts a superposition of the remainders of the divisions of r^n by N. This can be done in a small number of steps thanks to quantum parallelism, which allows us to process all the possible values of the exponent n in a single shot. The outcome is an entangled state of two registers, of which the second contains the remainder of r^n divided by N, when the first contains n. Now, as discussed earlier, it is impossible to extract from this state the *values* of the remainders. Any attempt to look into the second register forces the system to make a choice, whereby only one value of n and the corresponding remainder survive. But the values of the remainders are not needed. What is needed is the length of the cycle over which the remainders repeat themselves. This can be revealed in a few steps by applying a procedure which is the mathematical equivalent of passing a ray of light through a glass wedge to resolve it into its component colours—the colours of the rainbow.

To see how compelling the analogy is, remember that what our brain perceives as colour is the wavelength of the light, i.e. the spatial length of one cycle of an electromagnetic wave. The length of the cycle of remainders is like the 'wavelength' of that set of numbers. In this case, however, we know a priori that the set is periodic, which is like saying that we know our 'light' is of a single colour. But, as Newton first demonstrated, passing light of a single colour through a prism does not create additional colours: rather, it deflects the light by a single characteristic angle. Similarly, in the present case a spectral analysis of the second register will produce a single characteristic output, from which the order of the number r with respect to N can be inferred. And, from this, a prime factor of N can finally be extracted.

Kids' physics

Quantum mechanics has been called 'a young man's game'[11] and, in its early days (AD 1925), it was disparaged as kids' physics (Knaben-Physik), for many of its outstanding contributors were less than 30 years of age: Werner Heisenberg—24, Wolfgang Pauli—25, Paul Dirac—23, Pascual Jordan—23; with Erwin Schrödinger—37 and Max Born—42 already old guys. This is understandable. They needed the recklessness of youth to subvert an established way of thinking, to take the law into their own hands, to believe that the game they played with pencil and paper in a room was as real as the reality outside. The father founders of quantum mechanics were no fathers at all. They were *enfants terribles*, revelling in overlapping pools of truth and fantasy.

And yet, for all their faith in abstract mathematics and raw brain power, they ended up humanizing the discipline of theoretical physics. In the hands of classical physicists, nature had begun to resemble the mechanical nightingale of Andersen's tale—a clockwork that plays its tune regardless of whether the emperor is dead or alive. The quantum theoretical approach is less transparent, but more human-oriented than its classical counterpart. It is a way of thinking, not so much about nature itself, as on what nature will do for us in a given context, i.e. in a given experimental set-up. This way of thinking is rife with anthropic concepts such as uncertainty, probability, information. The state of the system is allowed to change abruptly when new information is acquired. The observer is allowed to become entangled with the object of the observation.

Once, during a walk, Einstein—an outspoken critic of the quantum theory—asked his friend Abraham Pais whether he believed that the Moon existed only when he looked at it.[12] To him the answer must have been obvious—as the fact that quantum theory was in conflict with it.

---

[11] J. Mehta, *The Birth of Quantum Mechanics*, CERN Report 76-10 (1976).

[12] A. Pais, *Subtle is the Lord: The Science and Life of Albert Einstein* (Oxford University Press, 2005).

Nine hundred years earlier the Persian poet and astronomer Omar Khayyam had already written a beautiful answer to Einstein's question (I quote the translation of Edward Fitzgerald):

> Ah, Moon of my Delight who know'st no wane
> The Moon of Heav'n is rising once again
> How oft hereafter rising shall she look
> Through this same Garden after me—in vain![13]

It is worth pausing for a moment to consider the image of the Moon rising majestically on the garden in which the two scientists once philosophized. But Khayyam saw that there were two Moons: 'The Moon of my Delight', subjective, personal, inextricably related to Man's hopes and destiny; and the 'Moon of Heaven', distant, aloof, bound to outlive Man by eons. This kind of duality is also in the heart of the quantum theory.

Quantum-like effects are commonplace in human psychology. We have all experienced situations in which, being undecided until the very last moment, our choice materialized only in the act of making it. Most of us have known the pain of loving and hating at the same time, and much of what we say is neither true nor false, but simultaneously true and false. Psychology, like quantum mechanics, thrives on the counterfactual: the mere awareness of a possibility affects our perception of the whole situation, as you can easily verify by walking on a narrow beam near the ground or up in the air between two skyscrapers. Similarly, the human response to situations is contextual: it depends on a multitude of facts, far and near, present and past, real or imagined. This context can be studied as a whole, but not taken apart. You cannot draw a box around a person and say: all that is important to this person's behaviour is here, within the box. And since quantum reality is so intimately shaped by human patterns of thought, is it so surprising that human behaviour is so often so well described by quantum mechanical metaphors?

[13] Omar Khayyam and Edward Fitzgerald, *The Rubaiyat of Omar Khayyam*, First edition.

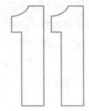

The spinning electron and other metaphors

A metaphor has two meanings for the reason,
but only one for the feeling,
so one who lives the world as a metaphor
could experience as a single thing
what has two entirely different meanings
in the eye of the world.

R. Musil, *The Confusions of Young Törless*

Doublets

It all began with a rainbow—the emission spectrum of sodium. Like other elements in the alkali metals family, sodium has an odd number of electrons, eleven, of which one is special: it is the most external, the most loosely bound, the first to be lost or shared in chemical reactions. While the other ten electrons huddle together around the nucleus, the eleventh lives a dangerous life at the margins of the electron cloud and can be compared, by way of metaphor, to the single electron of a hydrogen atom with an anomalously enlarged 'nucleus', which includes ten inner electrons.

It is the eleventh electron that controls the optical properties of sodium. For example, the yellow light of a sodium lamp is produced when the eleventh electron, normally in a state called '3s', is temporarily excited, by heating or by the passage of an electric current, to a higher energy state called '3p'. When the electron returns to the 3s state, the excess energy is released in the form of a photon of yellow light. All this was fairly well understood by spectroscopists of the early 1920s. But when they looked more closely at the colours of the spectrum, they realized that there was a *fine structure*: the yellow line was actually made up of two lines, two slightly different shades of yellow, too close to be distinguished by the naked human eye, but not so close that a good spectrometer could not tell them apart.

The mysterious splitting of the yellow line of sodium into two lines is the beginning of a fascinating tale of doublings interspersed with occasional halvings. The yellow doublet of sodium reveals a fundamental double-valuedness of the electron wave function—a double-valuedness that today we describe, metaphorically, as *spin*. In this chapter we will see how the double-valued electron wave function had to undergo a second doubling in order to be reconciled with the principle of relativity, and how the mind-boggling four-valuedness of the augmented electron wave function was brilliantly interpreted as the manifestation of a new particle—the anti-electron, now known as the *positron*. Finally we will see how these beautiful theoretical discoveries led to a description of elementary particles in terms of *quantum fields*, that is to say, fields of fields. Throughout this chapter the number 2 recurs as a mystical presence. Over and over, when we seem to be driven into a corner by unexpected observations, a providential doubling—a folding of the theoretical structure upon itself—restores our understanding, or the illusion thereof.

A good question with a better answer

The idea that the electron might be spinning like a fast top about its own axis had been around for a while, when some people began to suspect that it might be the key to the explanation of the fine structure in the light emission of atoms. The idea had first come up as a possible explanation of magnetism in matter (see Chapter 4). An electron has an electric charge, and a fast-spinning electron, similar to a microscopic current loop, would therefore produce a magnetic field. So a large assembly of electrons, all spinning together in the same direction, might well produce the strong magnetic field that is observed in natural magnets.

Remarkably, the same idea could explain the splitting of the energy levels of sodium. Picture the outer electron as a negatively charged spinning top going around the nucleus in a circular orbit, like the Earth around the sun. Now let us take an electron-centric view of this

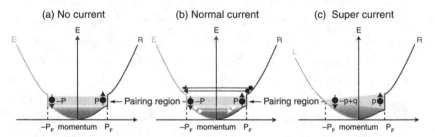

FIG. 77 A classical electron-centric model for the splitting of energy levels in a sodium atom. The outermost electron (the spinning particle at the centre of the figure) 'sees' the atomic nucleus and 10 core electrons circling around it. The energy is lower when the magnetic field produced by the spinning electron is parallel to the magnetic field produced by the atomic core (dark arrow), and higher when the two fields point in opposite directions.

miniature solar system. We put the spinning electron at the centre of the picture (Fig. 77) and let the atomic core (i.e. the nucleus with its following of ten core electrons) go around in a circle. The advantage of this 'Aristotelian' description is that we can clearly see that there are two current loops: the large outer loop described by the atomic core, and the small inner loop described by the spinning electron. When the magnetic fields created by these current loops point in the same direction, the system is 'relaxed' and its energy is low, but when they point in opposite directions the system is 'under stress', and its energy is higher. These two possibilities beautifully match the two energy levels that are seen in the split yellow line of sodium.

Unfortunately, this classical picture of the spin as a literal rotation is riddled with difficulties. A first difficulty comes from the magnitude of the effect. To account for the observed splitting of the yellow lines one needs a spin angular momentum of the order of Planck's constant, $\hbar = 10^{-34}$ joule-seconds. This seems rather small until one asks: how fast should the electron be spinning to attain such a value of the angular momentum? The answer depends on how large the electron is assumed to be. The modern theory bestows on the electron the rare honour of being a 'point-like particle', i.e. a particle of zero radius, but such a particle would not have any angular momentum in a classical sense. Within classical physics, an educated guess for the electron radius is about 10^{-15} m—comparable to the size of the atomic nucleus.[1] In order to have an angular momentum $\hbar$, an electron of this size would have to be turning about its own axis something like 10^{24} times per second, and the speed of a point at its surface would then be more than ten times the speed of light. Clearly this is in contradiction with the theory of relativity, underscoring the implausibility of classical physics at such small length scales.

---

[1] This estimate follows from the dubious assumption that the Einstein rest energy of the electron, $E = mc^2$, arises entirely from the electrostatic repulsion between different 'parts' of the electron.

A more fundamental problem arises when one tries to explain the fact that there are only two spectral lines in a doublet, and that they occur at definite wavelengths. To explain this, one must postulate that the magnitude of the spin angular momentum is fixed and that the particle can only spin clockwise or counterclockwise around the orbital axis (hence the double-valuedness). It is very difficult, if not impossible, to imagine a classical object that would behave in this way. The axis of a classical top can spin in any direction, and its angular momentum can have any value. Accordingly, one would expect to observe a continuous distribution of emitted light, not just two sharp lines. Even more worrisome is the fact that the angular momentum of a classical top can change in a collision with another object, but the spin of the electron does not change in spite of the electron's frequent collisions with other electrons, nuclei, etc.; it is a truly amazing, unstoppable top. The situation is reminiscent of the paradox of the constant size of atoms, which was so brilliantly solved by the quantum theory: the exploit would be repeated here in an even more fanciful way.

The concept of a quantized spin of the electron—an intrinsic spin angular momentum that could assume only the values $\hbar/2$ (counterclockwise or 'spin-up') or $-\hbar/2$ (clockwise or 'spin down') along *any* prescribed direction—was introduced by Dutch physicists Uhlenbeck and Goudsmit in 1926 (we will discuss the all-important factor 1/2— half—in the next section). Their speculation was supported by the results of a landmark experiment performed by Stern and Gerlach in 1922, which had clearly exposed a double-valuedness of the magnetization of silver atoms (the experiment is briefly described in the caption of Fig. 78).

Strangely enough, the bold and somewhat implausible suggestion of Uhlenbeck and Goudsmit was accepted almost immediately. A few quick discussions at railway stations, between one train and the next, were apparently sufficient to convince scientists of the calibre of Wolfgang Pauli and Werner Heisenberg. These young men were fully aware of the fact that the spin could not be given a literal interpretation in

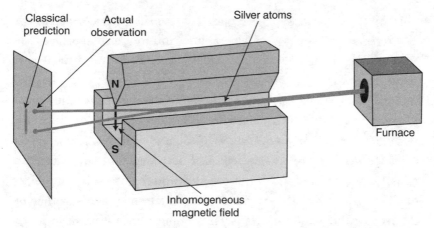

FIG. 78 In the crucial 1922 experiment by Otto Stern and Walter Gerlach a beam of silver atoms was passed through an inhomogeneous magnetic field. The classical expectation was that the atoms would be deflected up or down along the north–south line by variable random displacements. Instead all the atoms fell neatly on one or the other of two spots, as if their magnetization could have only one of two values, up or down. This observation is widely regarded as the experimental basis for the modern theory of the electron spin. In general, it is impossible to predict whether a spin-carrying particle will be deflected up or down by a Stern–Gerlach apparatus, but a particle that is deflected up enters a special state, called the 'spin-up state', which is sure to be deflected upwards by a second Stern–Gerlach apparatus. Similarly, a particle that is deflected down enters a 'spin–down state', which is sure to be deflected downwards by a second Stern–Gerlach apparatus.

terms of rotations, that it could not exist in the classical world. But to them the classical world was already a convention, a tradition inherited from the past, a fiction less real than the new world they were constructing. So what if the spin cannot exist, literally, as a rotation of a physical particle? What if the particle has shrunk to the point where there is nothing to rotate about? In the context of the quantum theory these objections are as idle as, in the context of religious faith, the contention that Jesus could not have walked on water. Just as for ideal circles and perfect squares, the non-existence of spin is the best guarantee of its existence. Soon, Pauli would show that spin is just a metaphor to evoke a certain type of behaviour of the electron wave function under rotations, and that the transformation properties implied by the half-integer spin require an 'exclusion principle'—no two electrons are allowed to be in the same quantum state—explaining the structure of the periodic table of elements, as well as the stability of matter in the universe.

What silenced many potential critics was the relative ease with which the introduction of spin provided an explanation for the optical spectra of atoms like sodium. And this was not just an explanation, but an explanation of exquisitely fine details. Let us return to the doublet in the emission spectrum of sodium. A question you might ask is: why only a doublet? Granted, each energy level splits into two levels, depending on the relative orientation of the orbital and spin rotation; but then this splitting should apply both to the initial and the final state of the electron, resulting in four, not just two lines—a quadruplet! Quantum mechanics has a very good answer to this question. The yellow lines of sodium arise from transitions between two quantum states, known in the jargon as the 3s and the 3p state. The 3p state has an orbital angular momentum—the electron goes around the nucleus, or the nucleus around the electron—and therefore gets split into two levels in the presence of spin. But the 3s state has zero orbital angular momentum—the electron does not rotate at all, it hovers around the nucleus in a spherical cloud supported by 'quantum pressure'—and for such a state there is no splitting. The reason is that the energy cannot depend on the absolute orientation of the spin, only on its orientation relative to the orbital angular momentum. But, there is no orbital angular momentum in the 3s state, hence no way of distinguishing between spin-up and spin-down, hence no splitting.

There is no better way to test a theory than to apply it to a scenario different from the one that initially prompted its development. The overall symmetry of an atom under rotations can be broken by the application of a magnetic field. This introduces a preferred direction in space—the direction of the magnetic field—and different orientations of the total angular momentum (orbital + spin) with respect to the magnetic field, will have different energies. We would then expect our

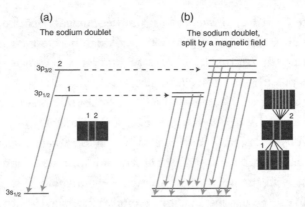

FIG. 79 How the two lines of the sodium doublet (a) split into ten lines (b) under the influence of a weak magnetic field. The two top levels split into two groups of four and two levels respectively, while the bottom level splits into two.

doublet to split into many more lines. How many? The answer is breathtaking: the upper line of the doublet splits into six lines and the lower one into four lines, for a total of ten lines. The diagram in Fig. 79 shows how this complex scenario arises from the splittings of the initial and final states of the outer electron in sodium.

The effect that I have just described played a key role in the development of quantum mechanics. Today it is still occasionally called the *anomalous Zeeman effect*, even though it is absolutely normal from the point of view of the quantum theory. Back in the 1920s, it was quite a challenge to understand how the magnetic field could produce different splittings in different groups of states (Fig. 79). In the absence of spin, all groups of levels should have split in the same manner, but then one should have observed only six lines—three from each component of the original doublet. The observation of ten rather than six lines suggested that somehow the orbital and spin components of the angular momentum respond to the magnetic field in different manners. More precisely, it looked as if the spin angular momentum interacted with the magnetic field twice as strongly as the orbital angular momentum— almost as if Nature wanted to make up for giving the electron only half a quantum of angular momentum $\hbar/2$ rather than the full quantum $\hbar$. This difference was a deep mystery as long as the spin was regarded as a literal rotation: but it came to be seen as perfectly natural as soon as the fetters of classical imagination were shaken off and the spin came to be viewed as an abstract feature of quantum particles.

This was the path chosen by Wolfgang Pauli and his contemporaries. They gave the electron an intrinsic magnetic moment[2] twice as large as predicted by a classical model, and went on to explain the anomalous Zeeman effect in full detail (the algebra is a little too complicated to describe). With the boldness of youth they shrugged off what initially looked like a fatal contradiction: namely, the fact that the doubled magnetic moment would produce too large a splitting between the lines of the sodium doublet. They were confident that better calculations would take care of the problem—and their optimism was rewarded. Shortly afterwards, Dirac developed an epochal theory which predicted the correct value of the magnetic moment *and* the spin-orbit splitting in a single shot, resolving the apparent contradiction between the anomalous Zeeman effect and the atomic doublets. We can glimpse here an extension of Machiavelli's motto: not only Fortune, but Nature too yields to the young and the bold.

Spinors

But if spin is not a real rotation, what is it then? Pauli's answer is a gem of fantastic precision: the spin is the anisotropy of the electron, it's a metaphorical way of saying that the wave function of the electron is not just a numerical field, like pressure or temperature, but has an internal structure which causes it to look different when viewed from different angles.

---

[2] 'Magnetic moment' is the quantity that controls the strength of the interaction of a current loop with a magnetic field. Similarly, the intrinsic magnetic moment of the electron controls the strength of the interaction between the electron and the magnetic field.

The idea, in itself, is not revolutionary. The electric field, for example, has a definite direction: it is a vector that can point towards you, away from you, or anywhere in between. In quantum mechanics, the wave function of the photon has a direction too; it is called polarization. This direction determines the probability that the photon will be absorbed or transmitted by a linear polarization filter. Superficially, the spin is to the electron what the polarization is to the photon: an internal orientation that determines the probability of the electron being deflected up or down in the Stern–Gerlach apparatus of Fig. 78. You will remember that in that experiment the spin-carrying particle passing through a strongly non-uniform magnetic field could be deflected either up or down, but never in between. Likewise, the photon could be either absorbed or transmitted by the 'polaroid' filter: again with no other possibility in between. The two mutually exclusive outcomes define two special internal states of the particle. For a photon, these are the **V**-state, which is sure to be transmitted through a vertically oriented 'polaroid', and the **H**-state, which will definitely go through a horizontally oriented 'polaroid'. For a spin the two states are the 'up'-state, which is sure to be deflected up in the Stern–Gerlach apparatus, and the 'down–state', which will certainly be deflected down.

So far everything looks the same, but upon closer inspection we realize that the behaviour of the spin under rotations is quite different from that of the polarization. A first clue to this is that the 'up' and 'down' orientations of a spin are related by a 180° rotation in ordinary space, while the horizontal and vertical polarizations of the photon are related by a 90° rotation. We will now see that the behaviour of the spin under rotation is, in a very precise sense, *intermediate* between that of a scalar quantity (temperature, pressure), which is invariant under rotations, and that of a vector quantity (electric field, polarization), which transforms like a step in space.

Scalars and vectors provide different representations of one and the same abstract group of transformations: rotations in space. In the scalar

representation all rotations are represented by identities, i.e. transformations that do not change the object on which they act. This is called the
'spin-0' representation. In the vector representation, rotations are represented by their action on oriented arrows. This is called the 'spin-1'
representation. But the wave function of an electron is *neither a scalar
nor a vector*. It belongs to a third class of objects called 'spinors', which
provide yet another representation of spatial rotations. The behaviour
of spinors under rotation is halfway between that of scalars and vectors:
the representation that these objects support is therefore aptly called the
spin-1/2 representation. Saying that the electron has spin 'one-half'
(in units of ℏ) is no more and no less than saying that its wave function
is a spinor.

Let me try to explain what a spinor is. I will do this by comparing and
contrasting it with a vector, namely the polarization of the photon. Both
can be represented by arrows, but you must bear in mind, now more
than ever, the relation between what you see in ordinary space and what
underlies it in the abstract space of quantum states, in which the spinor
lives. The relation between the visible and the invisible is one of the
main themes of this book, and it is the change in this relation which
makes all the difference between the spinor arrow and the vector arrow.

So let us begin with a **V**-photon, i.e. a photon that is guaranteed to
pass through a vertical polarizing filter, and its analogue, a spin-up
electron, i.e. an electron that is guaranteed to be deflected upwards by
a vertical Stern–Gerlach apparatus. Both states are represented by 'up'
arrows in the top diagrams of Fig. 80. These arrows indicate the orientation of the polarization or the spin in space—in what we might have
called 'real space', if 'real' were not such a loaded word. Now let us turn
the arrows clockwise. We are definitely changing the orientation of the
polarization and the spin. After a 90-degree rotation we pause to
evaluate the results. Although the polarization arrow and the spin
arrow are still pointing in the same direction, their significance in
terms of quantum states is vastly different. Here is where you must
imagine another reality behind the screen: to facilitate the effort, I have

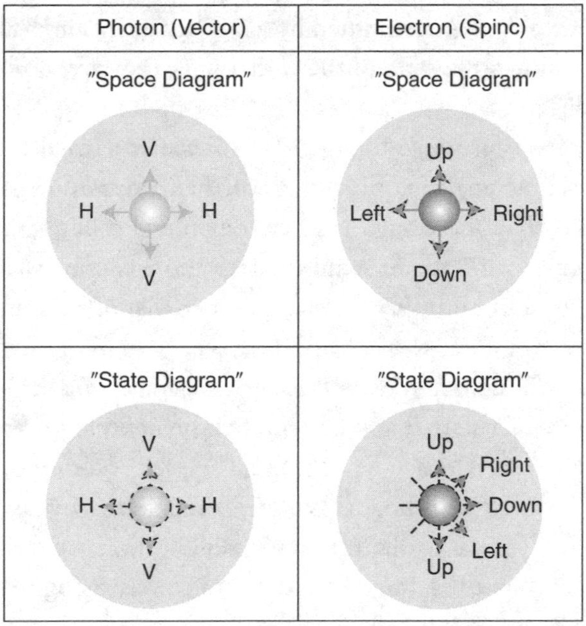

Photon (Vector)	Electron (Spinc)
"Space Diagram"	"Space Diagram"
"State Diagram"	"State Diagram"

FIG. 80 Left: The relation between the orientation of the polarization of a photon and the orientation of its 'state vector' in the abstract space of quantum states of the photon. Right: The corresponding relation between the orientation of the spin of an electron in real space and the orientation of its state vector in the space of quantum states (spinor space). Observe how a full 360° rotation in real space corresponds to only half a turn, i.e. a change of sign, in spinor space.

drawn in the lower part of Fig. 80 two more diagrams, which I call 'state diagrams', because they show what is going on in the space of quantum states. In the case of the photon, the state arrow coincides with the polarization arrow. So the state in which the polarization arrow is horizontal is an **H**-state, which is orthogonal to the initial **V**-state. For the electron, the situation is quite different. When the spin arrow in the space diagram points horizontally to the right, the corresponding quantum state is one in which the spin has equal probabilities of being deflected up or down by a vertical Stern–Gerlach apparatus. In this sense the state of the electron is like that of a D_I photon—a photon that is polarized at 45° with the horizontal and has therefore equal probabilities of being transmitted by a **V**-filter or by an **H**-filter. This is why I have drawn the state of the electron as an arrow pointing in the diagonal direction.

This observation suggests a general rule: whenever the spin arrow turns by a certain angle in ordinary space, the corresponding state arrow turns by only half that angle. You see how the idea of a 1/2 representation is dawning... To understand the idea more clearly let us continue our rotation until both the polarization arrow and the spin arrow are pointing straight down. Now the photon is back to its original **V**-state— a state that will definitely pass through a **V**-filter. But the spin is in a 'down state', i.e. a state that will definitely be deflected downwards by the vertical Stern–Gerlach apparatus. This state corresponds to the photon's **H**-state. It is 'orthogonal' to the spin up-state, just as the photon **H**-state is orthogonal to the photon **V**-state. In other words, the 'state arrow' for the spin has turned by only 90°, even though the physical orientation of the spin is turned by 180°!

At this point you can perhaps see the reason for the half-angle rule. The key point is that the two orthogonal states of a spinor, spin-up and spin-down, are connected by a 180° rotation in ordinary space, while the two orthogonal states of the polarization of a photon, vertical and horizontal, are connected by the usual 90° rotation. And this is why I said earlier that the spinor is a thing intermediate between a scalar and a vector. A scalar does not change at all under rotations. A vector, such as polarization, follows every rotation faithfully—its behaviour offering the most natural representation of a rotation. But a spinor follows spatial rotations half-heartedly, turning by only one half the angle of the spatial rotation.

An even stranger thing happens when we rotate our spin or polarization arrows a full turn: 360°. Now the space arrows are back where they started—pointing up—but the state arrow of the spin has turned only 180°, so we have got the negative of the initial spin state! This is very, very strange. Let me repeat it: under a rotation of 360° the arrow that represents the quantum state of the spin does not return to its original orientation, but to the negative of its original orientation.

It is of little comfort to think that the reversed arrow represents, physically, the same state as the original arrow. We learned in our childhood that a rotation by 360° is no rotation at all. And why should such a non-rotation require any change in our representation of the state of the spin? Before trying to answer this question, you should notice that two full turns in space, i.e. a rotation by 720°, do restore the state arrow to its original orientation, because half of 720° is 360°, which is one full turn in the space of states. So our puzzle can be restated as follows: why is it that two complete turns of a spinor are equivalent to no rotation, while a single turn leaves a trace—a change in the sign of the state arrow?

First of all, we must examine the origin of the prejudice that a rotation by 360° is no rotation at all. At first, this looks like unassailable truth. If I make a full 360° turn, I come back to where I started—no doubt about that. But the real question is not whether the physical states before and after the rotation are or are not the same (they are). The question is whether the *physical process* of turning an object by 360° is somehow different and distinguishable from taking no action at all.

Those who are old enough to have struggled in their youth with a telephone-cord will readily understand what I have in mind. Turn the handset by 360° and you will have a twisted cord, a testimony to the fact that there has been a 360° rotation. You can do the same thing more easily with a belt, holding one end firmly while turning the other—say the buckle end—by 360°. The resulting twist in the belt is evidence that one end has turned 360° relative to the other. This is expected. What is not so widely known is that if you twist the belt a second time, by turning the buckle end in the same direction another 360°, you actually untwist the belt! I must warn you that this is not immediately evident. After the second 360-degree twist the belt appears far more twisted than before. But if you move the buckle end, looping it around the belt itself, paying close attention to never turning your wrist—for that would be

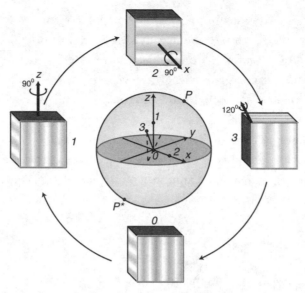

FIG. 81 The rotations of a three-dimensional object are in one-to-one correspondence with the points in the interior of a ball. For example, point 1 represents a rotation by 90° around the vertical (z) axis, point 2 represents a rotation by 90° about the x axis, and point 3 respresents a rotation by 120° about the diagonal of the cube. Point 0 represents no rotation at all. Two diametrically opposite points, P and P*, on the surface of the ball represent the same rotation.

cheating—you will see that somewhere along the path the twist drops out as if by magic, and you recover the straight belt. The lesson is that there is an essential difference between going around one full circle or two full circles (or, more generally, going around an odd or an even number of full circles): the former is an entangling process, which may leave a permanent signature in the environment, while the latter is truly equivalent to no rotation.

Figures 81 and 82 provide a more abstract explanation of the difference between one and two full turns. Each rotation is represented by a point in the interior or on the surface of a ball of radius 1. A point within the ball represents a counterclockwise rotation around an axis drawn from the ball's centre through the point in question. The rotation angle is 180° for a point on the surface of the ball, 90° for a point halfway

(a)   (b)   (c)

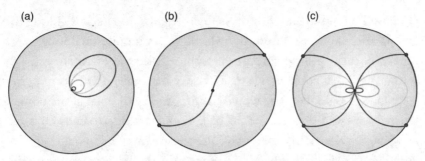

FIG. 82 Three closed paths in the 'ball of rotations' (Fig 81) represent three ways of changing the orientation of an object so that the initial and final orientations coincide. In (a) the object is taken around a loop, which can be continuously reduced to a single point at the centre of the ball. This is equivalent to no rotation. In (b) the object is taken around a full 360° rotation: the path pierces the surface of the ball and re-emerges at the diametrically opposite point. This loop *cannot* be continuously reduced to a single point. Finally, in (c) the object undergoes a 720° rotation: the path pierces the surface of the ball twice. Just as in (a), this loop *can* be continuously reduced to a point, showing the equivalence of a 720° rotation to no rotation.

between the centre and the surface of the ball, 0° for a point at the centre, and so on in proportion to the distance of the point to the centre of the ball. This is further illustrated in the caption of Fig. 81. With this convention, a point P on the surface of the ball represents a 180° rotation, and the diametrically opposite point P^* represents a 180° rotation about the reversed axis. But these two rotations are identical, as can be easily checked. Therefore the two diametrically opposite points describe the same rotation and must be regarded as one and the same point. We can exit the ball at any point on its surface, say P, and re-enter it immediately at the diametrically opposite point, P^*, as if the two points were contiguous.

The process of turning an object by 360° can be viewed as the tracing of a closed path that starts at the centre of the ball, goes out along a radius, reaches the surface, smoothly crosses over to the diametrically opposite point, and returns to the centre (see Fig 82(b)). This path is closed, yet it is essentially different from a closed loop that is entirely contained within the interior of the ball (Fig. 82(a)). The closed loop of Fig. 82(a) could be continuously reduced to an infinitesimal loop near the centre of the ball, which represents the identity, i.e. no rotation at all. But the path that corresponds to a 360° rotation in Fig. 82(b) cannot be so reduced. To do that, you would have to first cut it open at the points where it crosses the surface of the ball, then bend the two ends inwards and glue them back together. Only after doing this could you begin the continuous reduction towards the centre of the ball. But cutting and gluing are not legal operations: resorting to them would be like solving the Rubik's cube by disassembling it, then reassembling it in the right order. So there is an essential difference between a 360° rotation and no rotation.

But, you say, what about turning an object by 720°? The closed path that describes this process pierces the surface of the ball twice before returning to the centre. Why don't we run here into the same difficulty that prevented us from reducing a 360° rotation to no rotation? It turns out that we don't, because the presence of two points at which our path pierces the surface of the ball gives us a way to effect the reduction without breaking the rules of the game. The idea is to deform the path in such a way that the two points of intersection with the surface of the ball come closer and closer until they merge together, allowing the path to smoothly enter the ball (Fig. 82(c)). From this point on, the path becomes a closed loop entirely contained within the ball and can be continuously shrunk to the single central point that corresponds to no rotation. This is why a 720° rotation is equivalent to no rotation, while a 360° rotation is not.

After this brief excursion into a field that mathematicians call 'topology', let us return to the properties of spinors and vectors. Having established that a 720° rotation is truly equivalent to doing nothing, we can deduce that a 360° rotation can either do nothing or change the sign of the quantum state. This is because a 720° rotation can be viewed as the combination of two 360° rotations. If each of the two 360° rotations produces a change in sign, then the 720° rotation produces no change in sign because the product of two minus signs is a plus sign. And, of course, if each 360° rotation produces no change in sign, then a 720° rotation produces no change in sign either. No other possibility exists, given that the state must be invariant under a 720° rotation. Therefore every legitimate representation of rotations must be either an 'integer' representation, which leaves its objects invariant under 360° rotations, or a 'half-integer' representation, whose objects change sign under 360° rotations. Vectors belong to the first class, spinors to the second. Both represent the same abstract group of rotations, and both occur in nature, as we have seen from the examples of the polarization of the photon and the spin of the electron.

The exclusion principle

As for the electron, it is hard to overestimate the consequences of that minus sign in a 360° rotation of the spin. Consider a pair of electrons. Now interchange them. Inevitably, one of the two spins turns by 360° relative to the other, as you can see in Fig. 83; this multiplies the quantum state of the pair by a minus sign. So the wave function of any two-electron system must change sign when the two electrons are interchanged. But this implies that you can never put two electrons in the same quantum state—a fact known as *Pauli's exclusion principle*.

You can easily see why it must be so. Suppose you could do just that—put two electrons in the same quantum state: then the combined state of the pair would be completely symmetric upon exchange of the electrons. Why? Because the two electrons are absolutely identical twins: there is no way to tell them apart, and hence no way to tell whether they have been interchanged or not, *as long as they are in the same quantum state*. But such a symmetric state, by the very definition of 'symmetric', cannot change sign when the two particles are interchanged. So we have two contradictory requirements: on the one hand the combined state of the two electrons *must change sign* upon interchange; on the other hand, *it cannot change sign*. The only way out of

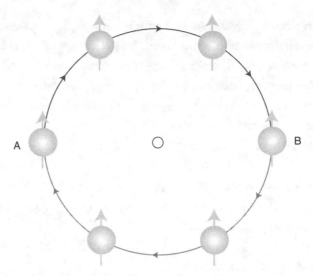

FIG. 83 Interchanging two particles A and B entails a 180° rotation around the midpoint O of the segment that joins A to B—or a 360° rotation of one particle relative to the other.

the contradiction is to recognize that *two electrons can never be in the same state*. If an electron is in a certain state of motion around the nucleus, then a second electron will be allowed in the same state only on condition of having opposite spin orientation, and a third one will not be allowed at all, but will be forced to occupy a different state, even if it costs considerable energy to do so.

It is by reason of this iron-like necessity that electrons build up the staggering variety of chemical elements. We have already encountered hydrogen, the most abundant element in the universe—engaging, reactive, always eager to enter chemical bonds. Helium, with one more electron, is in a completely different class—a 'noble gas', aloof, chemically inert. Lithium, with three electrons, again a mercurial personality, father of many structures and compounds. Carbon, with its six electrons, the world's most accomplished networker, responsible for the hardness of diamond, the softness of graphite, the flexibility of membranes... Each additional electron brings in new individuality, new chemistry, the marvellous variety of molecules that congregate into matter, matter that becomes life, life that supports consciousness. And all this stemming from the Pauli exclusion principle, which is a direct consequence of the half-integer value of the ineffable spin of the electron.

Dirac's void

By 1926, the spinning electron, with the backing of Pauli's exclusion principle, was firmly established as the most important particle of matter. At last, one could understand the periodic table of elements as a systematic filling of shells of electronic states in the atom. But the solution of a problem always creates new problems: the very same progress that had established the leading role of the electron in chemistry also touched off a chain of events that was going to undermine the very concept of 'quantum particle', on which quantum mechanics had been built.

The decisive step in this direction was taken by Paul Dirac whom we first met in Chapter 6 as the champion of 'invariants', and then again in Chapter 8, trying to stay cool after sighting the connection between Poisson brackets and commutators. More than anybody else Dirac contributed to bringing together the two great physical theories of his time: quantum mechanics and special relativity. The tension between these two pinnacles of theoretical physics had been mounting as quantum mechanics bloomed into maturity. It was not just a matter of the difference between a new and an old way of thinking: the key equations of the new quantum theory contradicted the letter of the theory of relativity.

In order to be 'relativistically correct' a theory should treat time and space even-handedly, i.e. as two faces of the same coin, which indeed they are. Unfortunately, quantum mechanics posits a sharp difference between time and space—a difference that has its root in the very concept of particle. Time is a number that labels different instants, without being associated with any specific property of the particle. Space is different. Surely you can define a position in space without any reference to particles, and in this sense 'position' is just a label for different points. But, in quantum mechanics this label is also one of many possible outcomes of an experiment designed to determine the position of the particle, and this clearly *is* a physical property. Given this fundamental difference, it is perhaps not surprising that the most important equation of quantum mechanics, the *Schrödinger equation*, which determines the time evolution of the wave function ψ for a particle, grossly breaks the equivalence of space and time. The Schrödinger equation has the schematic form

$$iD_t\psi = -D_xD_x\psi$$

where D_t is a mathematical operator that computes the rates of change in time, while D_x computes the rate of variation in space, and i is the imaginary unit. Even without knowing the mathematics of it, you can see that space (x) and time (t) enter the equation un-symmetrically: the spatial operator D_x appears twice, whereas the time operator D_t appears only once. For this reason the Schrödinger equation fails to satisfy the principle of relativity, i.e. to keep the same form in all reference frames that move with constant velocity relative to each other.

Dirac's resolution of this fundamental difficulty is a striking example of how one can simultaneously succeed and fail: win a battle and lose it at the same time. The first challenge was to find the correct relativistic generalization of the Schrödinger equation. This should still have the form $iD_t\psi$ = something-proportional-to-ψ, because, according to general principles of mechanics, the state represented by ψ contains all the information needed to determine its own future—in particular, its own rate of change in time. But now we also have the principle of relativity telling us that the something-proportional-to-ψ on the right-hand side of the equation must contain only *one* operator D_x if the equivalence of space and time is to be preserved. Terms like $D_xD_x\psi$ or higher powers of

D_x are not allowed, as they would break this equivalence. Finally, and this is crucial, the something-proportional-to-ψ must arise from the application of the energy operator H to the wave function. This is because the energy is, in mechanics, the generator of the time evolution (see Chapter 8), and indeed the Schrödinger equation is a way of saying just that.

These requirements are highly problematic. In quantum mechanics the operator D_x corresponds to the momentum of the particle, p, so that the above discussion boils down to saying that the energy operator H must contain p linearly, i.e. p as a single power, and not p^2 or p^3 ... But the energy of a non-relativistic particle is proportional to p^2 and the energy of a relativistic particle is even worse: something like the square root of $p^2 + m^2$, where m is the mass. Obviously, it is hard to see how, barring witchcraft or wizardry, the square root of $p^2 + m^2$ can be recast in a form that contains a single power of p, as demanded by the principle of relativity.

I fancy that Dirac first saw the possibility of performing the magic when, after a lot of fooling around with the maths, he realized that the square root of $p^2 + m^2$ can indeed be linear in p, provided that one is willing to increase the dimension of the space of states on which p operates. The wave function ψ that appears in the Schrödinger equation is a single complex number, one for each value of x and t. What if we upgrade it to *two* complex numbers, together forming a spinor as Pauli had suggested? The reason why this might work is that in the enlarged space of two-component wave functions there are some non-trivial operations which, applied twice in sequence, yield the 'identity operation', i.e. the operation that leaves everything unchanged. Consider, for example, the operation X that interchanges the two components of the wave function: if you apply this operation twice, i.e. if you interchange the components of the wave function twice, you obviously come back to the initial wave function. The fancy way to express this fact is to write $X^2 = 1$, where X^2 is X times X (i.e. X applied twice), and 1 denotes the identity operation. The same can be said of an operation Z that changes the sign of the second component of the wave function without touching the first: apply it twice in a row and you don't get any change, therefore we have $Z^2 = 1$. Furthermore, if you apply first X and then Z you obtain the negative of what you would obtain by applying first Z and then X (try it!) This means that we have $ZX = -XZ$ or, equivalently, $ZX + XZ = 0$. Making use of these results you may be able to verify that $(pX + mZ) \times (pX + mZ) = p^2 + m^2$. But this means that $pX + mZ$ is the 'square root' of $p^2 + m^2$, and yet it contains only the first power of p... So we see that in a space of two-component wave functions it is mathematically possible to express the energy by a formula that contains only p and not p^2—but will it really work? Will it really lead to a consistent and relativistically correct theory?

In a hypothetical world with only one spatial dimension, the two-component wave function described above would have worked satisfactorily, but not in our three-dimensional world. In three dimensions we need at least two more operations like X and Z to accommodate two more mutually perpendicular directions of the momentum. Dirac discovered that the simplest scheme that could accommodate these additional operations required not two, but *four-component* wave functions. He thus introduced a new type of wave function consisting of two spinors inextricably bound together to form a *double spinor*. And so, with his double spinor firmly in hand, Dirac was able to construct a beautiful equation that treated space and time symmetrically, that was invariant under relativistic transformations of reference frame, that predicted an intrinsic angular momentum of ℏ/2 (the spin), and the splitting of the atomic doublets, and the anomalous Zeeman effect—all correct and all in one shot. An extraordinary success by all measures... except for one small detail:

> As whence the sun begins his reflection
> Shipwracking storms and direful thunders break
> So from that spring whence comfort seemed to come
> Discomfort swells.[3]

[3] Shakespeare, *Macbeth*, Act I, Scene ii.

No sooner had genius broken into the fortress of relativistic quantum mechanics than reality struck back with a new and dreadful menace: the negative-energy solutions. For the Dirac equation has an infinity of negative-energy solutions: for every solution with positive energy E there is another solution with negative energy $-$E and there is no limit on how large the energy can be on either side. Mathematically this happens because the square root of $p^2 + m^2$, from which the energy is obtained, can have two values with opposite signs, just as the square root of 4 can be $+$ 2 or $-$ 2 (both numbers square up to 4). And so you have not only two possible orientations of the spin, but also two possible signs of the energy, which 'explains', a posteriori, why the wave function has four rather than two components.

The trouble with negative-energy solutions (or, more accurately, with arbitrarily large negative energies) is that they make no sense physically. Leaders of the capitalistic economy might welcome their existence for providing the final solution to their energy needs: they could dump any kind of matter into negative-energy states and pull out fresh positive energy. However, to a scientist this likely means that the universe is unstable and will quickly sink into the bottomless pit of negative-energy states.

To be fair, negative energies might occur also in classical relativity and for the same reason (there are two square roots of $p^2 + m^2$, with opposite signs), but there they are immediately dismissed as a mathematical artefact. This can be safely done because positive and negative energy states are well separated and never mix. Why can't we do the same in quantum mechanics?

The reason is subtle and profound, hinging on the very definition of particle. Remember that a particle is something that can be localized, something that always appears *whole* at one point—never at two points at the same time. This is true both in classical and in quantum mechanics. But in quantum mechanics you can form localized states only by superimposing infinitely many states of different energy. And in relativistic quantum mechanics this superposition must necessarily include

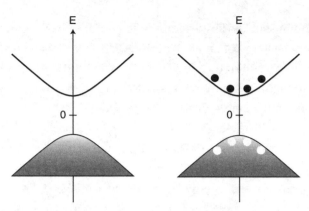

FIG. 84 Left: the Dirac Void. All the negative-energy states—the Dirac sea—are occupied, while the positive energy states are empty. Right: a few electrons have been promoted to positive-energy states, leaving vacancies in the negative-energy sea. The vacancies are interpreted as anti-electrons, better known as positrons.

negative-energy states. There is no way of fully localizing a particle without giving it a finite probability of being in a negative-energy state. Negative energies, with their following of unnatural consequences, are not only possible, but absolutely necessary to the existence of a relativistic quantum particle.

> Dismayed not this
> Our captains, Dirac and Pauli? Yes:
> As sparrows, eagles, or the hare, the lion.[4]

Look at the solution they came up with. The negative-energy states, says Dirac, are there, but they are already occupied by electrons—one electron in each state, as shown in Fig. 84—and never mind the fact that the Universe should then explode as a result of the repulsion between the electrons. For we are going to call this ocean of unseen particles 'the Void', the dark cosmic background, the Nothingness out of which the world was created. The Void is the state of minimum energy. Relative to the Void every other state must have positive energy. Why? Because of Pauli's exclusion principle, of course. No two electrons can occupy the

---

[4] Shakespeare, *Macbeth*, Act I, Scene ii.

same quantum state. So any attempt to lower the energy of the Void by decreasing the energy of an electron will fail, because the lower-energy state is already occupied by an electron and cannot be doubly occupied.

Here is something else we can do. Pull an electron out of the Void and put it in one of the empty positive energy states. Now we have created not one, but two particles (Fig. 84). First, we have promoted one of the unseen electrons from the Void into the real world, where it appears as a positive-energy electron; second, we have left a 'hole' back in the Void— a hole that has positive energy, because a missing negative energy is a positive energy. The hole represents a particle no less real than the electron. For example, a hole can jump from one state to another within the Dirac Void: all it takes is that one of the electrons of the Dirac Void fill the hole, leaving a vacancy in the state whence it came. This process mirrors the jump of a real electron from one state to another in the positive energy sector, but takes place entirely within the negative energy sector. There is another important difference. When a hole goes from point A to point B, what really happens is that an electron—one of the unseen ones—moves from B to A. Saying that a negative charge— the charge of the unseen electron—has moved from B to A, is the same as saying that a positive charge has moved from A to B. In short, the hole behaves, for all practical purposes, like an electron of positive charge, and is in every respect as real or as fictional as the electron. This new particle is now called a positron—and was the first particle of *antimatter* ever predicted to exist.[5]

Other exciting things can happen. For example, a positron meets an electron and the two particles annihilate each other, emitting electro-magnetic radiation. The electron falls into the hole, which the positron metaphorically stands for. Afterwards there is no electron and no hole,

---

[5] Dirac himself did not attach great significance to the experimental observation of the positron in 1932—nor did he encourage experimentalists to hunt for the new particle. To him, getting the equations right was the real achievement.

and the lost energy is carried away by a photon of the electromagnetic field.

What about the problem of pinpointing a particle at a certain position in space? I said that it could not be done without making use of negative energy states. Initially those states seemed crazy and unphysical, but now we see that they have an interpretation as positron states. What does this tell us about the nature of localized particles? Well, it tells us that there is no such a thing as a strictly localized particle: as with all ideas of physics, that of a localized particle makes sense only *within limits*.

It should be noted that the background electrons in the Void occupy states of definite momentum, which are, *ipso facto*, completely delocalized in space (see Chapter 9). As soon as we attempt to confine one of the electrons by applying a force field we induce a 'stress' in the Void surrounding the particle. The Void is no longer uniform and, as a result, some *virtual* electron–positron pairs begin to appear. Virtual they are because they are not fully expressed in reality, just as the polarization of a photon is not fully expressed before the observation that reveals it, but, as the particle is confined to a smaller and smaller region of space, the stress increases, until, at a critical length scale of the order of 10^{-13} m, the Void breaks down and *real* electron–positron pairs begin to be produced spontaneously. They draw energy out of the very same force field which we had hoped to use as a tool to pinpoint the particle. The shocking conclusion is that the number of particles is no longer a well–defined quantity! We start from one particle, but, as we attempt to localize it, we generate many. Strictly speaking, it makes no sense to say 'here is a particle' because the quantum state always includes the possibility of more and more particles (electron–positron pairs) popping out of the Void.

And so a single particle does not exist in the literal sense, but only as a metaphor to describe a far more complex state of infinitely many particles. Yet this complex state looks and acts like a single particle if one takes a somewhat lenient view of it, i.e. if one does not pry too sharply into its inner structure. The technical name for such a state is a *quasiparticle*, by which we mean something that looks like a particle on a sufficiently large scale, but reveals its complex nature when we put it under a sufficiently powerful microscope.

The lesson is that there are no true particles in nature, only quasi-particles. You can see now why I said that Dirac won and lost his battle. He did get the relativistically correct wave equation, but he also found that it no longer described a single quantum particle.

Quantum field theory

After reading the last few paragraphs, no one should ever think again that physicists are any less imaginative than novelists or poets. But art takes more than imagination. Form is equally important. The French painter and sculptor Edgar Degas complained to the poet Stefan Mallarmé: 'Yours is a hellish craft. I can't manage to say what I want, and yet I am full of ideas...' 'Dear Degas—replied Mallarmé—one does not make poetry with ideas, but with words'.[6]

Good form usually means simplicity and economy of thought. Dirac's idea of a sea of invisible particles, useful as it was to grasp the concept of antimatter, strikes us as implausible, anti-economical, baroque. Why fill the universe with an infinite number of electrons to explain the existence of a single positron? Why burden the Void with an infinite load of matter? Why expend many means when a few may suffice?

This criticism is the starting point of *quantum field theory*. Instead of talking of individual electrons and positrons, the quantum field theorist imagines a universal electron–positron field, meaning an infinite set of quantum mechanical operators which, at every point in space and time, can create or destroy an electron or a positron. The global state of the quantum field is described by a wave function at each point of space and time—hence the earlier suggestion that a quantum field is a 'field of fields'. This wave

---

[6] The exchange is reported by Paul Valèry in *Poetry and Abstract Thought*.

function determines the probability of finding an electron or a positron here or there, now or then. But these electrons and positrons are just the foam, *maya*—they are things that can be created at a point and destroyed at another. Only the field is permanent. The field has its own Void, namely the state in which there are no electrons or positrons. But this does not mean that the field itself is absent in the Void. True, it is zero on average, but it is continuously present in the form of random deviations from the average, which are called fluctuations. One way to picture these fluctuations is to imagine pairs of particles—electron–positron pairs—popping out of the Void, hanging there for a few instants, and going back to the Void before anything or anyone has had the chance to detect their presence. These particles, which live unobserved for the short time of a fluctuation, are called *virtual* to distinguish them from real particles, which can definitely be observed. The Void is continuously seething with such invisible bubbles. Another field—the electromagnetic field—can latch onto the fluctuations of the electron–positron field and produce real particles out of the Void. For example, a photon of the electromagnetic field can be converted into a real electron–positron pair, which can later recombine, creating a real photon. Particles come and go, but the field is eternal.

A most attractive feature of quantum field theory is that it naturally explains why all electrons are identical. This becomes completely obvious when you think of them as local manifestations of a universal field. On the downside, the solution of the equations of motion for the quantum field is far more difficult than the solution of the equation of motion for a particle. Even the relatively simple theory known as *quantum electrodynamics*, which includes only two fields—the electron–positron field and the electromagnetic field—has not been solved exactly. Yes, there are powerful techniques, which enable physicists to calculate and predict facts and numbers with spectacular accuracy, but these techniques include mathematical tricks to dispose of infinite quantities that pop up in the calculations. *Renormalization theory* is the name for this trickery. Dirac complained about it, and even Feynman,

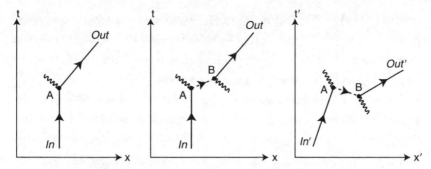

FIG. 85 Left: an electron interacts with the electromagnetic field at space-time point A and makes a direct transition from the initial to the final state. Middle: the electron now goes from the initial to the final state passing through two interactions with the electromagnetic field, and the creation of a virtual electron (dashed line) in between. Right: the same process as seen from a reference frame that is travelling to the left. The virtual electron propagates backwards in time, meaning that we have a virtual positron propagating forwards in time. One man's virtual particles are another man's virtual antiparticles.

who won the Nobel Prize for his contributions to quantum electro-dynamics, declared in his own characteristic style: 'I think that the renormalization theory is simply a way to sweep the difficulties of the divergences [i.e. the infinities] of electrodynamics under the rug'.[7]

In spite of these difficulties, quantum field theory does provide many deep and beautiful insights. I will give you only one example, namely the reason for the existence of antimatter.[8] Consider a process in which an electron jumps ('makes a transition') from an initial state to a final state by interacting with an electromagnetic field. In the quantum field theoretical description, it is not the electron that goes from one state to another, but rather it is the field that destroys an electron in the initial state and creates one in the final state. In a sense we are back to Zeno's old idea of motion as illusion: first, you have a particle in this state; later, you have a particle in that state—but you cannot tell how it went from

[7] R. P. Feynman—Nobel Prize Lecture, 1965.

[8] See R. P. Feynman, *Elementary Particles and the Laws of Physics* (Cambridge University Press), of which the following paragraphs are a summary.

one to the other. The left panel of Fig. 85 shows the most direct route from the initial to the final state: the field destroys the incoming electron at space-time point A and immediately creates the outgoing one at the same point. But this is only the simplest route.

The middle panel of Fig. 85 shows a more complicated process in which the electron created at space-time point A propagates to space-time point B, where it is replaced by the outgoing particle. Many other possible routes—actually, infinitely many—are described by diagrams of increasing complexity, but for our purposes it is sufficient to concentrate on this one. The intermediate particle, which propagates between A and B, is an example of a virtual particle, and should be clearly distinguished from the real particles, incoming and outgoing, which can be observed—the first coming from the remote past, the second going into the indefinite future. The virtual particle has some strange features that make it quite different from a real particle. For example, it is not subject to the common rule that its speed must be less than the speed of light. The events A and B may be arbitrarily far apart in space—in fact so far apart that no physical signal could possibly connect them in the intervening time $t_B - t_A$, yet the route remains 'legal' and does contribute to the total transition probability—not much, but not zero either. This paradox epitomizes the breakdown of the individual particle concept. There is no relation of cause and effect between the creation of the virtual electron in A and its destruction in B. Rather, the creation of an electron in A instantaneously alters the probability of an electron being destroyed in B. This is another example of the 'spooky action at a distance' that Einstein decried. Here it strikes again, but does not contradict relativity, because the electron that is destroyed in B is not the same electron that was created in A.

The non-causal propagation of virtual particles is the ultimate reason why antimatter is not only possible, but necessary. To understand why this is so, consider again the process depicted in the middle panel of Fig. 85, but now imagine yourself moving with constant velocity to the left. Surely you remember the two busy pedestrians of Fig. 33 (p. 115) who, walking in opposite directions, had different views of past and future: for one walker, the cataclysmic stellar explosion was still in the future, while for the other it was already in the past. Those two representations of reality could peacefully coexist because neither walker had any way of knowing anything about the impending explosion.

Well, something similar is going on here. To an observer who is moving to the left, the events A and B may appear in reverse temporal order, i.e. the time of event B may be earlier than the time of event A. The right panel of Fig. 85 shows the process as seen by this observer. The difference is striking: the virtual particle is going back in time. It was hard enough to explain how the virtual particle could travel faster than light. Now we must explain how it can travel backwards in time!

But the principle of relativity assures us that the situation cannot be as absurd as it seems. We only need to look at it from a different point of view. The key to this different point of view is that a virtual electron running backwards in time represents a virtual positron moving forwards in time. Armed with this insight let us try and read the right panel of Fig. 85 in the proper manner, starting from the time of event B. At that time an electron–positron pair pops out of the Void—a result of the interaction with the ever-present electromagnetic field; the electron immediately emerges in the final state while the positron is virtual. However, both the electron and the positron propagate forwards in time. At the later time of event A, the virtual positron meets the incoming electron and the two annihilate each other. This sequence of events makes perfect sense, and the final observable result is still the same as in the left panel, even though the description of the process, at the virtual level, is completely different. The principle of relativity says that both observers are right; they are simply using different words to describe the same reality. Which means that if electrons exist so must positrons exist too, because we now clearly see that one man's virtual electrons can be another man's virtual positrons! The collusion between relativity and quantum mechanics is essential to the argument. Without relativity we would not be able to change virtual particles into virtual antiparticles. But without quantum mechanics we would not have virtual particles or antiparticles to begin with: there would be only real particles, which can never transform into real antiparticles.

Superconductivity

Pages ago I promised to say something about superconductivity—the phenomenon in which certain metals lose, at low temperature, their electrical resistance. It is fitting to conclude this chapter by honouring that promise, if nothing else because the tale will highlight the twin themes of spin and Pauli's exclusion principle, and bring forth a new and more credible incarnation of Dirac's Void.

So let us start at the beginning and let us ask why, in ordinary metals, the flow of an electric current is such an unnatural state—a state that must be enforced by an external power source (e.g. a battery), and that quickly dies off if that source is turned off. Earlier I asked you to picture a metal as a sea of electrons moving chaotically in all directions and responding with the utmost sensitivity to the force exerted by an electric field. That was a classical description. It also holds in quantum mechanics with two important differences. First, because of Pauli's exclusion principle, no more than two electrons can be in the same state of motion, and two electrons in the same state of motion must have opposite spins. Second, the state of motion of an electron is characterized by its momentum, which, in a system of finite size (say a large box) can only assume certain discrete values in each direction.[9] Now let us say that there are N electrons in the metal and we want to distribute them among different states of motion in such a way that the system has the lowest possible energy. Because the energy of an electron grows with increasing magnitude of its momentum, we will occupy the states

---

[9] As the size of the box tends to infinity the allowed values become so closely spaced as to be indistinguishable from a continuum.

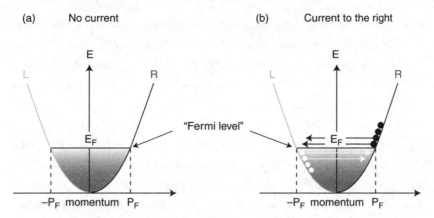

FIG. 86 (a) The state of least energy for electrons in a normal metal is analogous to the Dirac Void: all the states of momentum k smaller than the Fermi momentum k_F are occupied. Since there are equal numbers of right-moving and left-moving electrons, the net current in this state is zero. (b) In order to create an electronic current we transfer some electrons from left- to right-moving states. The resulting current is unstable because right-moving electrons can switch back to left-moving states, at no energy cost. In the absence of external forces this process continues until the number of electrons on the two sides is equal, and the net current is zero.

of lowest momentum first, and will consider higher momentum states only when the lower ones are fully occupied.

Let us suppose, for simplicity, that the electrons can only move to the right or to the left along a straight line. The first two electrons will go in the state of zero momentum, which carries zero energy. The next four electrons will go in the two states of lowest non-vanishing momentum, one moving to the right and one moving to the left. Four more electrons will go into two states of slightly higher momentum and energy, and so on, until all the electrons have been accommodated. Obviously, there is a huge number of electrons to be accommodated (10^{23} or so per cubic centimetre), and when we finish we realize that we have compactly filled states with momenta comprised between $-p_F$ and $+p_F$ and energies lower than E_F. The special momentum p_F, which separates the occupied states from the empty ones, is known as the *Fermi momentum*, shown in Fig. 86(a), and the corresponding energy E_F of an electron at the top of the distribution is known as the *Fermi energy*. This compact occupation scheme produces the state of least energy—also known as the ground-state—of the electrons in the metal. Notice that there is no net current in the ground-state because, for each electron moving to the right, there is another electron moving to the left with the same speed: the two opposite currents neutralize each other.

The similarity between the ground-state of the metal and the Dirac Void is truly striking: both are states of minimum energy populated by an immense number of electrons and stabilized by Pauli's exclusion principle.[10] In both cases an electron can be pulled out of the occupied sea leaving behind a positively charged hole (Fig. 86(b)). There is a crucial difference, however. In a metal, an electron–hole pair can be created with an infinitesimally small expenditure of energy. By contrast, the creation of an electron–positron pair out of the Dirac Void requires an energy no less than $2mc^2$, the sum of the rest energies of the electron and the positron. We will see that a minimum energy also appears in a superconductor.

Now, if the ground-state of a metal carries no current, how do we get a current to flow, say, to the right? The obvious solution is to increase the number of electrons that move to the right and correspondingly reduce the number of electrons that move to the left. The new situation is shown in Fig. 86(b), and you can immediately see that the new state is an *excited state*, i.e. it has higher energy than the ground state. Why so? Well, you can certainly lower the energy of this state by transferring some of the top right-moving electrons back to one of the empty left-moving states. In other words, the state is unstable against processes that transfer electrons from the right- to the left-moving group. Such processes are continually occurring, due to collisions with impurities, or with vibrating atomic cores in the background. The essential point, for

---

[10] The peaceful coexistence of many electrons in a metal is made possible by the fact that their negative charge is neutralized, on average, by the positive charge of the nuclei of the atoms that constitute the metal.

our purposes, is that the current-carrying state *has both occupied and unoccupied states at the same energy*. Therefore an electron can collide with an impurity, or with a vibrating nucleus, and switch from a right- to a left-moving state at no energy cost. Of course, the reverse can also happen. But there are more right-movers than left-movers, so these random collisions will cause a systematic transfer of electrons from the right-moving to the left-moving side. As a result of this, the current will steadily decay to zero, unless there is an external power source to maintain it. This is the basic mechanism of electrical resistance in normal metals.

The question now is: what is so special about superconductors, which allows them to elude the decay process described in the previous paragraph? And when I say 'elude' I really mean it, because the current in a superconductor is virtually eternal: it can flow for millions of years, in spite of impurities, atomic vibrations, etc., as long as the temperature remains sufficiently low.

An ingenious and beautiful theory of superconductivity was devised by Bardeen, Cooper, and Schrieffer (BCS) some fifty years ago and is justly considered one of the highest achievements of the physics of the twentieth century. BCS observed that the interaction between electrons at energies close to the Fermi level (the energy E_F in Fig. 86) is often attractive rather than repulsive.[11] When attraction prevails, the plain free-electron ground state depicted in Fig. 86 breaks up in favour of a different kind of state, which may carry a current without the drawback of having both occupied and unoccupied states at the same energy.

Figure 87 illustrates the main features of the state proposed by BCS. Far from the Fermi level this state is indistinguishable from the normal state of electrons in a metal: the electrons occupy momentum states subject only to the restrictions imposed by the Pauli exclusion principle. But, in

[11] The origin of this subtle effect is in the tiny deformation of the crystal brought about by the passage of an electron: the deformation carries a positive electric charge which, in turn, attracts a second electron.

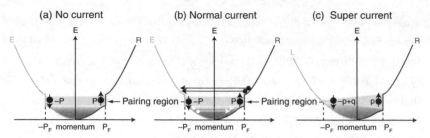

FIG. 87 Three states of a superconductor. In the first (a) we have pairing between electrons of opposite momentum and spin, and no net current. In (b) we have a normal current created by breaking some of the pairs and transferring some electrons from left- to right-moving states. The resulting current is unstable. In (c) the electrons are still organized in pairs, but the momenta of each pair do not add up to zero. This state carries a current, but no pair is broken. Even though its energy is higher that the energy of the ground state, there is no viable mechanism for it to decay. The current is a *supercurrent*, and is eternal.

the vicinity of the Fermi level a new kind of correlation appears between the occupancies of different momentum states. More precisely, a spin-up electron will occupy a state of momentum p if and only if a spin-down electron occupies the state of opposite momentum −p. In practice, this means that, if you find a spin-up electron in a state of momentum p, then you can be sure, without checking, that there is a spin-down electron in the state of momentum −p. And if you do *not* find a spin-up electron in the state of momentum p, then you can safely bet that there is no spin-down electron in the state of momentum −p. You may protest that this correlation is already present is the normal ground-state of Fig. 86. But the essential difference is that, in the normal state you can, with negligible expenditure of energy, promote one of the top electrons to a slightly higher momentum state, where it has no partner of opposite momentum. So, the pairing correlation that you discern in Fig. 86 is accidental and easily destroyed by small perturbations. In the superconducting state, however, the pairing condition cannot be broken without a finite expenditure of energy: it is a robust feature of the ground state.

At variance with a normal metal, it can never happen in a superconductor that a state of momentum p, spin up, is occupied, while the state of momentum $-$ p, spin down, is empty. Well, I take it back: it can happen, but at a price—an energy price—as if each member of the pair had to give up something valuable, some kind of 'life insurance', in order to live as a single particle. Think about it again, and you may discern an analogy with electron–positron pairs in the Dirac Void. Let us say that the ground-state of the superconductor, with its ocean of nicely paired electrons, is the Void. The breaking of a pair in the superconductor corresponds to the creation of an electron–positron pair out of the Void. For example, the pair (p↑, $-$ p↓) might be broken by pulling out the electron p↑ and putting it in an initially empty state p′↑. What you get is an un-paired electron in p′↑ and an un-paired hole in p↑. The essential point is that you have to pay a minimum energy price 2Δ to create the pair of excitations (Fig. 87), just as you had to supply a minimum energy $2mc^2$ to create an electron–positron pair out of the Dirac Void. So, here is the sharp difference between a superconducting metal and a normal metal. In a normal metal, electrons can be re-arranged at zero energy cost, but in a superconductor there is a finite energy price to be paid for breaking an electron pair.

What does all this have to do with superconductivity? After all, the paired ground state, which I have just described, still carries no current, because the members of each pair move in opposite directions. This is a good place to demystify a common metaphor which explains super-conductivity as a form of compassionate solidarity between electrons. The electrons of a superconductor—so goes the metaphor—are like soldiers marching in tight ranks against the enemy: when one stumbles—which is inevitable in view of the asperities of the terrain, fatigue, snipers' fire, etc.—the others support him and put him back on his feet. So the current goes on forever. Unfortunately, the facts are quite different. Not only do the soldiers move in different directions, but the members of each pair—the inseparable buddies that should be caring for each other—run in opposite directions, more keen on escaping from their partner than on meeting him. One can, of course, create an ordinary electric current by breaking some pairs and thus generating some electrons and an equal number of holes as shown in Fig. 87(b), but such a current would decay by the usual mechanisms, under the action of impurities and imperfections.

A true supercurrent—a current that never decays—can be established by enforcing the following rule. Whenever the spin-up member of a pair moves to the right, with momentum p, its spin-down mate must move to the left with a momentum that is not exactly $-p$ but $-p + q$, so that the pair as a whole (or, better, that abstract point known as the centre of mass of the pair) travels with momentum q to the right. Now there is a net current flowing, as shown in Fig. 87(c), but the situation looks very artificial and likely to be destroyed by impurities in much the same way as the normal current of Fig. 86(b) ... or not?

Take a second look. In Fig. 86(b) an electron can collide with an impurity and change its direction of motion without changing its energy. This is also possible in Fig. 87(b), but not any more in Fig. 87(c). The reason is that the current-carrying state, just as the ground-state, does not contain any un-paired electrons that might change their direction of motion at zero energy cost. You see, if an electron wanted to reverse its direction of motion, changing momentum from $+ p$ to $- p$, then it would have to break a pair, i.e. create an electron and a hole, which cost a minimum energy Δ apiece. Where could that energy be coming from? Impurities or nuclear vibrations cannot supply it, for more or less the same reason that prevents us from extracting energy from the waves of the sea. There is no way for the electrons to pull themselves out of the 'social contract' in which they got stuck. Like prisoners led to an execution they have nowhere to go, short of breaking their chains and fleeing—two feats for which they do not have enough energy. So the cooperative behaviour of the electrons is not a game of falling and being put back on your feet by your nearest comrade, but one of unswerving obedience to a marching order that was perhaps unanimously chosen at some point, but now must be accepted by everybody, even by those who had nothing to do with choosing it.

I am deliberately using sociological metaphors. The only way for the electrons to stop flowing would be to do so collectively: in other words, to renegotiate the terms of their association into pairs. This is not absolutely impossible, but it is extremely unlikely to happen without an external intervention. One could compare the process to the task of rewriting the constitution of a country, only much more arduous. Historically, constitutions change every few decades, but supercurrents live as long as the universe.

Collective states—of which the superconducting state is an example— have this common feature: the individual constituents (electrons in this case) stand in an egalitarian relation to the order that governs their behaviour: they are simultaneously its subjects, its creators, and its enforcers. Like the citizens of a classless society, they do not obey an external authority; they obey only themselves. Through the practice of civil disobedience a society can transcend its own laws, which is the key to ethical progress. But it takes enormous courage to be one of the initiators of the transformation, since one is almost sure to be destroyed before gathering sufficient consensus. Electrons, trapped in the deep valley of their superconducting state, have no choice but 'going with the flow'. We humans, organized in societies, have a moral duty to question the established laws, to conceive new possibilities, to imagine new valleys beyond the mountain ridges that limit our field of vision. Hence the importance of fantasy and courage.

12

Margarita

Just as the Moon in her orbit around the Earth keeps a side of herself permanently averted from us, so every successful theory hides a body of implicit assumptions, unstated limits, narrowly escaped contradictions, which will be seen only by those who dare the journey to the other side. Only these adventurers will have a sense of the *inner frontier*— by which I mean, the frontier that limits the depth of our understanding, just as the outer frontier limits the extent of our knowledge. Researchers at the inner frontier question the very foundations of science. They ask whether they have been deceived by their teachers or are unwittingly deceiving their own students. They ask, for example, whether quantum mechanics and relativity are as universally valid as we have been taught, or are they limiting cases, shadows of more comprehensive theories? And are the laws of physics as permanent as we make them to be, or do they too, perhaps, evolve in time?

Physicists often talk of a 'final theory' which would unify everything there is to know about the Laws of Nature, and thus bring about the death of physics as we know it. Should we believe this? Imagine a world in which all the basic laws of physics had been discovered and written down in a single gold-plated Book—the true and veritable Bible of Physics. So what? What is the value of such a Book if there is no human mind to grasp it, no human mind to be set on fire by it? The Book has no life until you, the reader, open it. But you will not learn much from such a Book unless you are able to recreate the subject in your mind, which means, in practice, writing your personal copy of the Book on your own paper and in your own handwriting. In the end it's only what we carry deep in our heart that constitutes real knowledge.

Picture yourself as a newcomer to the discipline. At first you accept concepts and definitions with humility and faith, awed by the authority of the great scientists who have written the Book. You are unsure of yourself in the unfamiliar territory. Every small irregularity of the terrain trips you. Then, as you gain confidence and skill, you begin to understand the topography of the field, you move quickly and confidently along its pathways, you don't let yourself be distracted or intimidated by small and big contradictions that always loom on the horizon like dark clouds, you see trouble when it is still far away and skilfully manoeuvre so that it never crosses your path: at this stage you really are a steamroller, ready to solve problem after problem...And then, just as you thought you had it all figured out, something terrible happens: a simple question, possibly asked by an eight-year-old—a question to which you have no answer. And not because it takes knowledge that you don't have, but because it takes *understanding* that you don't have.

Then you withdraw in solitude. By now you are so familiar with your subject that you carry it around wherever you go, like a backpack permanently hung on your shoulders. You reconsider all that you know—what have you been missing? You fancied yourself a giant, but now you realize that even if you were to add six feet to your stature you would still be a dwarf. And then, all of a sudden, comes a flash of insight. The horizon recedes: a new circle of thoughts comes into view.

In all likelihood, this will be just a personal achievement, one that will make you a deeper scientist or a better teacher, but will not carry fame into the outer world. The 'new' insight had probably been in plain sight for a long time. Perhaps the writers of the Book knew of it. Perhaps you were the only one who didn't! Once in a long while, if history is well disposed and if you have been chosen for such a destiny, your insight may open up a new way of thinking—a new arena in which others can show their prowess. Then you will have made your big contribution to science, and a new edition of the Book will have to be written.

There are no final truths at the frontier, no fixtures—only, an inexhaustible activity that creates and continually destroys its own creations. Yet, none of the fixtures we admire in the outer sphere of life would be possible without the work that goes on in this inner region. It seems to me that all accumulated knowledge is of little worth in comparison to the fire in the belly of the individual scientist. We see some of the greatest human achievements exposed to public admiration in academies, museums, opera houses. Great works of art, when put on display, appear necessary and definitive. But a phantom inhabits the dungeons of the opera house, and he is an arsonist. Nothing is definitive. Admiration is a form of misunderstanding.

In Chapter 2, I introduced the hero of *The Master and Margarita*—the Master of fantastic precision. With unerring imagination he guessed every detail of scenes he had never witnessed—the interrogation of Jesus by Pilate, the crucifixion of Christ. A consummate artist at that, he didn't have the strength to complete the task he had initiated. At the beginning of the novel the Master is a weak man who has yielded to formidable opposition and disavowed his own work. He is a broken man, who spends his days whispering words which nobody will listen to. To his rescue comes Margarita.

If the Master has the artistry, Margarita has the vital force to bring it to fruition. If the Master is prisoner of an institution, Margarita is an agent of subversion. Empowered by the Devil, she represents the victorious spirit which purifies our culture of prejudices and conventions. She is, I think, the ultimate hero of the scientific enterprise.

Perhaps the most dramatic moment of the novel occurs at the beginning of the 21st chapter when Margarita—until then the respectable wife of a wealthy businessman—is transformed into a beautiful witch. To the magic words of 'I'm invisible! I'm invisible and free!', she flies out of the window, leaving behind clothes, jewels, and respectability. Naked as the naked truth can be, she flies on a broom, invisible, in the night sky of Moscow. She roams over the city and the countryside. She punishes the enemies of the Master. She purifies herself in water and blood. She hosts the Grand Ball of Satan. She forgives old sins and settles old accounts. She rejoins the Master and recovers his manuscript. And then 'the crimson full moon rose up to meet them from behind the edge of the forest, all illusions vanished and the magical mutable clothing fell into the swamp and drowned in the mist' (p. 321, 32).

The Master and Margarita, *The Man Without Qualities*, *The Little Prince*: three masterworks that accompanied us through this journey, yet are never mentioned in the same breath. What do they have in common? Ideally placed at sunset, they are, all three, tales of transcendence, of departure from the material world. This book too is about transcending the world, about reaching out to an invisible reality that can never be fully comprehended. It is about the search for the truth having more value than the truth itself. It is about Columbus taking off for his final journey:

> Shadowy vast shapes smile through the air and sky,
> And on the distant waves sail countless ships,
> And anthems in new tongues I hear saluting me.[1]

---

[1] Walt Whitman, *Prayer of Columbus*.

PHOTOGRAPHIC ACKNOWLEDGEMENTS

© Mary Evans Picture Library/Alamy: fig. 67 (Cardano); National Gallery of Art, Washington D. C. Photo: © The Art Gallery Collection/Alamy: fig. 5; Israel Museum, Jerusalem, Gift of Harry Torczyner, New York. © ADAGP, Paris, and DACS, London 2010. Photo: © The Bridgeman Art Library: fig. 1; © The Field Museum, Chicago (CK24_1T): fig. 4; © James Steidl/iStockphoto: fig. 21; © Ben van der Zee/iStockphoto: fig. 64; © Drazen Vukelic/iStockphoto: fig. 16; © The Palace Museum, Beijing: fig. 20.

INDEX